oekom, naturally

This book puts a true slice of sustainability into your hands. In buying it, you are supporting production methods based on strict environmental standards.

In producing our books, we ...
... use mineral-oil free inks,
... do not wrap them in plastic packaging
... finance climate and biodiversity projects, and
... print them in Germany, thus guaranteeing short transport distances.

Further information is available at www.natuerlich-oekom.de and #natürlichoekom

oekom, naturally · sustainable since 1989 ·

Bibliographic information for the Deutsche Nationalbibliothek: the Deutsche Nationalbibliothek lists this publication in the Deutsche Nationalbibliografie; detailed bibliographic data are available on the Internet at www.dnb.de.

published 2025 oekom verlag, München
oekom – Gesellschaft für ökologische Kommunikation mbH
Goethestraße 28, 80336 München
+49 89 544184 – 200
info@oekom.de

Layout and Typesetting: oekom verlag
Cover Design: Laura Denke, oekom verlag
Cover Illustration: © Stefan Krauter
Print: Elanders Waiblingen GmbH, Waiblingen

ISBN 978-3-98726-501-3
E-ISBN 978-3-98726-497-9
https://doi.org/10.14512/9783987264979

RECYCLED
Papier aus Recyclingmaterial
FSC® C016368

Christine Freitag, Katrin Temmen,
Burkhard Hehenkamp, Tobias Klaus and
Stefan Krauter (Hrsg.)

Sustainable Energy Development in East Africa

Interdisciplinary Approaches

oekom

Content

Abstract 11

Table of Abbreviations and Formula Symbols 13

1. Introduction 17

Christine Freitag, Katrin Temmen

2. Aspects of (Failing) Sustainability in Energy Development: A Community-Owned Microgrid Case Study 21

Teddy Mangeni, Henry Asiimwe, Paul Bogere, Lillian Donna Namujju, Ibrahim Mwammenywa, Godiana Philipo, Josephine Nakato Kakande

2.1 Engineering Aspects 23

2.2 Economic Aspects 25

2.3 Social Scientific Aspects 26

2.4 Didactics of Technology Aspects 32

2.5 Conclusion 36

2.6 References 37

3. Improving Sustainable Energy Engineering 41

3.1 Josephine Nakato Kakande, Godiana Philipo, Tobias Klaus: Using and Storing Sustainable Energy 41

3.1.1 Introduction 41

3.1.2 Context and Background 43

3.1.3 Renewable Energy Sources in East Africa 44

3.1.4 East African Energy Mix 51

3.1.5 Generation and Storage Options 55

3.1.6 Typical Energy Use Cases and Load Profiles in East Africa 55
3.1.7 Recommendation 58
3.1.8 Conclusion 59
3.1.9 References 59
3.2 Stefan Trittler, Joachim Böcker, Henrik Bode: Designing and Constructing Microgrids for Local Communities 63
3.2.1 Important Design Considerations for Microgrids 64
3.2.2 Development and Applications of Power Electronics 71
3.2.3 Implementation of a Microgrid 78
3.2.4 Safety Aspects 81
3.2.5 Realization of Microgrid Demonstrator in Lwak, Kenya 83
3.3 Josephine Nakato Kakande, Stefan Krauter, Godiana Philipo: Demand-Side Management 94
3.3.1 Introduction 94
3.3.2 Typical DSM Approaches 95
3.3.3 DSM Case Study in Lwak, Kenya 96
3.3.4 Conclusion and Outlook 102
3.3.5 References 103
3.4 Ibrahim Mwammenywa, Ulrich Hilleringmann: Towards Sustainable Microgrids Development: Role of Wireless Sensor Systems and Signal Processing—Case of a Rural Community Microgrid in Tanzania 105
3.4.1 Introduction 105
3.4.1 Communication Media in Electric Grids 107
3.4.3 Implementation of Load Monitoring and Control Using LoRa and Fuzzy Logic in a Community Owned Microgrid 109
3.4.4 Conclusion 117
3.4.5 References 118

4. Societal Perspectives 119
4.1 Tobias Klaus: Energy Development in East Africa 119
4.1.1 Historical Context and Post-Colonial Energy Paradigms 119
4.1.2 Evolution of Energy Supply 119
4.1.3 Development of Energy Demand 121

4.1.4 Political and Social Institutions: Drivers and Barriers 121
4.1.5 Social and Political Problems 123
4.1.6 Conclusion 124
4.1.7 References 124
4.2 Socio-Cultural and Institutional Economic Analyses of Microgrids 126
4.2.1 Burkhard Hehenkamp: Introduction 126
4.2.2 Henrietta Acquah-Swanzy: Do Institutional Factors Affect Households' Willingness to Pay for Microgrid Electricity in Tanzania? 127
4.2.3 Lillian Donna Namujju, Burkhard Hehenkamp, Teddy Mangeni: Technical Capacity and Hierarchical Dynamics in Community-led Microgrids 138
4.2.4 Irene F. Ngoti: How Ownership Matters for Sustainable Implementation of Microgrids 151
4.2.5 Lillian Donna Namujju, Burkhard Hehenkamp: Sustaining Off-Grid Energy through Informal Governance: Network Evidence from a Community Microgrid in Tanzania 159
4.2.6 Burkhard Hehenkamp: Conclusion—Sustainability and Performance Outcomes in Microgrid Operations 170
4.2.7 References 171
4.3 Stakeholder Conflicts in Energy Communities 176
4.3.1 Teddy Mangeni: Introduction 176
4.3.2 Teddy Mangeni, Henry Asiimwe, Paul Bogere, Ibrahim Mwammenywa, Josephine Nakato Kakande: Interdisciplinary Conflict Perspectives: »Traffic Light System Conflicts« 177
4.3.3 Henrietta Acquah-Swanzy: The Impact of Nudging and Centralized Control on Sustainability in Flat-Rate Microgrids—A Common-Pool Resource Perspective 185
4.3.4 Irene F. Ngoti: Ownership and Conflict in Community-Based Microgrids: Insights from Rural Tanzania 190

4.3.5 Teddy Mangeni: Participatory Conflict Analysis and Community Development 198
4.3.6 References 207
4.4 Henry Asiimwe: Gender Discourses in Energy Development 210
4.4.1 Introduction 210
4.4.2 Literature on Gender Discourses and Barriers to Equitable Participation in Renewable Energy Processes 212
4.4.3 Local Discourses on Women's Participation: Insights from East Africa 215
4.4.4 Conclusion 221
4.4.5 References 222

5. Educational Perspectives 225
5.1 James Musana: Education for Sustainable Development: East African Perspectives 225
5.1.1 Introduction 225
5.1.2 Key Initiatives and Strategies for ESD in East Africa 227
5.1.3 Perspectives of Education for Sustainable Development in East Africa 229
5.1.4 Conclusion 233
5.1.5 References 233
5.2 Paul Bogere, Henrik Bode, Katrin Temmen: The Use of Demonstrator Grids in Teaching Microgrid Development 236
5.2.1 Introduction 236
5.2.2 The Lwak Microgrid as a Local Multiplicator 237
5.2.3 Don Bosco Grid 238
5.2.4 Microgrid Demonstrators at German Universities 241
5.2.5 References 241
5.3 Paul Bogere, Katrin Temmen: Developing a Virtual and Interactive Microgrids Learning Environment 243
5.3.1 The Role of VIMLE in Contributing to Sustainability 244
5.3.2 VIMLE as a Learning Tool—Who are the Stakeholders? 245
5.3.3 VIMLE Conceptual Model 247
5.3.4 VIMLE Implementation Challenges and Opportunities 255

5.3.5 Conclusion 256
5.3.6 References 257
5.4 Teddy Mangeni, Paul Bogere, Henry Asiimwe, Christine Freitag: Approaches to Community Learning in Microgrid Development 260
5.4.1 Introduction 260
5.4.2 Media and Methods of Community Learning in Microgrid Communities 261
5.4.3 Communities as a Resource for Learning 263
5.4.4 Gender Discourses and Community Learning: Challenges and Pathways to Inclusive Participation 265
5.4.5 References 269

6. Prospects: Knowledge Distribution and Educational Formats in Years to Come 273

Christine Freitag, Katrin Temmen

List of Figures 275
List of Tables 279
Editors and Authors 281

Abstract

This book explores sustainable energy development in East Africa from an interdisciplinary perspective. At its core lies the interaction between technological innovation and social, economic, cultural, and educational dimensions, which are crucial for the success of local and regional energy transitions. The contributions are based on the outcomes of the A:RT-D Grids project (2020–2024), an international partnership between East African and German universities, enterprises, and NGOs.

The chapters address practice-oriented questions on energy generation, storage, and use, with a particular focus on microgrids in rural regions. Alongside engineering approaches, the book examines societal dynamics such as acceptance, gender equity, institutional frameworks, and stakeholder conflicts within energy communities. In addition, the role of education for sustainable development is emphasized, highlighting its importance for capacity-building, knowledge sharing, and the long-term establishment of sustainable practices.

The book demonstrates that sustainable energy development extends far beyond the deployment of renewable technologies: it requires collaboration across disciplines, cultures, and institutions. The insights gained contribute to bridging scientific concepts with local realities and to developing solutions that are ecologically viable, socially just, and economically feasible. In doing so, this volume contributes to the broader discourse on global sustainability goals and regional transformation processes in East Africa.

Table of Abbreviations and Formula Symbols

Abbreviation	Designation
A:RT-D Grids	Africa: Research and Teaching for Development Grids
A	Ampere (Electrical Current)
Ah	Ampere-hour (Unit of Charge)
AI	Artificial Intelligence
AMS	Accounts Management System
AR	Augmented Reality
ARFMS	Assessment, Results, and Feedback Management System
ATP	Ability to Pay
BMS	Battery Management System
CA	Constructive Alignment
CMS	Content Management System
CPR	Common-pool Resources
CuPayment	How Much Households Currently Pay for Electricity in a Month
DC	Direct Current
DEO	District Energy Office
DOD	Depth of Discharge
DSM	Demand-Side Management
DT	Didactics of Technology
DTTF	Democratized Technical Training Framework
EA	East Africa
ELP	Electronic Learning Platforms
ESD	Education for Sustainable Development
GT	Graduate Training

Abbreviation	Designation
HR	Human Resource
ICMS	Information and Communication Management System
IGO	Intergovernmental Organization
IIoT	Industrial Internet of Things
ILO	Intended Learning Outcomes
kW	Kilowatt (Power)
LFP	Lithium Iron Phosphate
LMS	Learning Management System
MS	Management System
MVC	Model View Controller
NIE	New Institutional Economics
NMC	Lithium Nickel Manganese Cobalt Oxide
O&M	Operation and Maintenance
OOPMS	Other Online Platforms Management System
PESTEL	Policy, Economic, Social, Technological, Environmental, and Legal
PUE	Productive Use of Energy
PV	Photovoltaic
REA	Rural Energy Agency
REC	Regional Economic Community
R-Labs MS	Remote Laboratories Management System
SACMS	Synchronous, Asynchronous, and Cooperative Classrooms Management System
SD	Sustainable Development
SDG	Sustainable Development Goal
SHS	Solar Home System
SNA	Social Network Analysis
SoC	Relative State of Charge of an Energy Storage (e.g. a Battery)
SSA	Sub-Sahara Africa
STEM	Science, Technology, Engineering, and Mathematics
STS	Socio-Technical System
TEAMS	Tests, Exams, and Assignments Management System

Abbreviation	Designation
TLS	Smart Metering—Traffic Light System
V	Volt (Voltage)
VEC	Village Electricity Committee
VIMLE	Virtual and Interactive Microgrids Learning Environment
V-Labs	Virtual VIMLE Laboratories
VR	Virtual Reality
VTC	Vocational Training Centre
W	Watt (Unit for Power)
WASH	Water, Sanitation, and Hygiene
WTP	Willingness to Pay

Christine Freitag, Katrin Temmen

1. Introduction

This book is about sustainable energy development in rural East Africa. The contributions to the book have been discussed and delivered by experts from East Africa and Germany. All the contributors are connected having worked together in an international partnership programme for sustainable innovation that was funded by the German Ministry for Education and Research from 2020 to 2024. The A:RT-D Grids project (https://bmbf-client.de/en/projects/art-d-grids) linked universities, enterprises, and NGOs mainly in East Africa and Germany to address energy transformation by joined research and teaching. Both research and teaching have had practical and problem-oriented aims. The will to help develop new and improved ways of electrifying regions not yet covered by any electrical grid connection brought together experts from different fields. Even though everybody in the field seems to address the need for technical innovations to be linked to social realities, it often seems that technical and social sciences are viewed within a hierarchy, with technical innovation coming first. This book may go to show that sustainability in energy development is not merely about the use of renewable energies but about true collaboration between the technical and the social, in science as well as in the reality of energy development. As we focus on energy provision in local rural communities, this book covers engineering and social sciences in their different linkages to the energy sector, paying special attention to microgrids and local communities.

The book has been designed and set up very much along the lines of what we found to be worth sharing and teaching, especially during our three large graduate trainings in 2021, 2022, and 2023, which included hundreds of participants from all parts of the world. So, while the reader will find »classical« content associated with different academic disciplines, much of this is also linked to field realities and stories reconstructed by empirical research in local communities over the years. These realities have changed our disciplinary teaching in several ways. Not only has our teaching become adapted knowledge, but, in many ways, the deep dives into East African realities by so many people with so many different backgrounds have changed our academic knowledge inventory. Empirical challenges have had a major impact on theory.

Changes in our ideas of relevance were also induced by the participants of the graduate trainings, most of them students, some of them scientists, others technicians, secondary school teachers or college lecturers. For all graduate trainings, our evaluations show that participants demand practical relevance and problem-oriented teaching and are full of praise where this seems to have been successful. This, too, the reader will find in these pages. It may apply to single topics like »microgrid design« (chapter 3.2) or to the book as a whole. For instance, a microgrid case study is introduced in chapter 2, and this is then used as the problem and case base in many of the following chapters, highlighting some key aspects of focus.

The subtitle of this book calls for »interdisciplinary approaches«. At least three quarters of our training participants came from the engineering sector, and thus we would expect to have readers from this sector who want to learn to think outside their purely technical box. But we also had quite a few individuals from the social sciences with research interests in sociology, economics, as well as education and didactics. They wanted to learn technical basics and discuss the social realities of sustainable development. They made us as researchers and developers, authors and editors join in smaller and bigger teams to design the chapters and weave a common thread. While technicians sought updates on technical and other relevant insights that could improve their work, energy enterprises were keen to know all about field research and its relevance for energy policies. So, this is exactly what we hope to provide with this book.

One big challenge in our trainings was accessibility. Many people told us how difficult it was to secure a refund for travel expenses or to pay for mobile data, or to get time off to join our trainings. Many told us there was a far higher educational demand than could be covered. As such, this book is also a way of making knowledge more accessible to many, providing open access information on highly relevant and research-based facts, and, through this, promoting broader and deeper assistance in developing the sustainable energy sector.

Teddy Mangeni, Henry Asiimwe, Paul Bogere, Lillian Donna Namujju, Ibrahim Mwammenywa, Godiana Philipo, Josephine Nakato Kakande

2. Aspects of (Failing) Sustainability in Energy Development: A Community-Owned Microgrid Case Study

More than three-quarters of all the people in the world without access to electricity, and more than a third who lack access to clean cooking fuels and/or technologies, are in Africa. Most of the 600 million without electricity and the 970 million who relied on harmful fuels for cooking in 2021 were Sub-Saharan Africans. Almost half of them are concentrated in five countries: the Democratic Republic of the Congo (DRC), Ethiopia, Nigeria, Tanzania, and Uganda (International Energy Agency, 2023, p. 107). Despite the great strides made in electrification access (with 91 % of the world's population now having access to electricity), 10 million more people still lacked access to electricity as of 2022. Sub-Saharan Africa's share of the global deficit stands at 83.3 % with 473 million people in rural areas lacking access to electricity. The remaining un-electrified population is more challenging to serve because much of it is remote and low-income. Decentralized solutions (including stand-alone systems in remote and sparsely settled areas) offer a cost-competitive alternative to grid expansion and can be rapidly deployed to meet levels of demand too low to justify grid investments. In addition, productive uses in rural communities, such as solar water pumps, refrigerators, agro-processing machinery, and a wide range of equipment for micro enterprises, contribute to socio-economic development and improve quality of life. Achieving Sustainable Development Goal (SDG) 7.1 by 2030 is possible only by deploying a combination of grid, mini-grid, and stand-alone off-grid solutions that leverage the faster deployment of distributed renewables to efficiently meet demand. National and regional electrification programmes using public funding to unlock private co-investment at scale can bring these solutions to life (IEA et al., 2024, p. 11). The co-existence of different electrification options with a high share of microgrids and stand-alone systems—primarily Solar Home Systems (SHS)—helps to boost competition and accelerate electrification while enhancing reliability and resilience to climate and security risks in SSA (Sub-Saharan Africa) (International Energy Agency, 2023, p. 109).

Despite the underscored urgency of the task of transforming Africa's energy sector and the need to transition to cleaner, sustainable, and accessible energy sources, which cannot be emphasized enough, the renewable energy transition is a formidable challenge. African nations grapple with many interconnected obstacles ranging from inadequate energy infrastructure and financial constraints to policy (Agoundedemba et al., 2023, p. 2). While solar microgrid deployment can have enormous benefits for rural inhabitants in SSA, as with other technological developments, deploying them is not without barriers (Nuru et al., 2021, p. 2). Come Zebra et al. (2021), in their review of off-grid electrification systems in developing countries, note that despite the considerable growth in the number of microgrids implemented across the globe, there are still barriers that hinder the successful implementation of these systems in developing regions. In their study, they analyse the barriers using the PESTEL framework to examine the policy, economic, social, technological, environmental, and legal (PESTEL) aspects that can aid decision-making in addressing and overcoming the barriers to the implementation of these systems (Come Zebra et al., 2021, p. 18).

The A:RT-D (Africa: Research and Teaching for Development Grids) project recognizes not just that the sustainable operation of infrastructure systems depends on sound technology, but also that social and economic factors influence sustainable operation. The project, therefore, in its common idea for »Resilient Modular Grids for Increased Security and Sustainability of Electricity Supply«, undertook an interdisciplinary research approach using the mini-grids of East Africa to investigate the various aspects of microgrid sustainability (Krauter et al., 2019, p. 7). The empirical data collected from the field captured the reality of the visions of microgrid projects vis-à-vis the reality of microgrid project implementation. The various interdisciplinary findings, as presented and discussed in the subsequent chapters of this book, informed knowledge-based interventions and recommendations. In this section, the reality of a failing microgrid is presented and discussed through an interdisciplinary lens that takes in perspectives from engineering, economics, social sciences, and didactics of technology. Using the case of a rural community-managed microgrid in Tanzania, the presentation explores the actuality of the process of microgrid project implementation and its outcome on the state and sustainability of the project. Data in the Tanzanian case was collected over several data collection phases. While the discussions in this chapter present findings from the initial state of a community-owned microgrid, subsequent chapters consider evolutions of the Tanzanian case based on its interactions with research and time.

2.1 Engineering Aspects

Sustainable mini-grid development in Africa requires technically robust systems that can withstand local environmental conditions, operate reliably, and be maintained with limited resources. However, challenges such as inverter-induced harmonics and limited local technical capacity highlight the need for better power quality control and community training. The use of simplified components and local technician involvement demonstrate the importance of maintainability for long-term sustainability. These observations align with broader findings discussed in the case study as described.

The present case study is of a village located in a rural region of Tanzania. It has a community-owned Solar PV Microgrid. The microgrid is situated at the village's centre. It serves about 60 households and shops, spanning a radius of about 1 km from the microgrid plant. The Solar PV microgrid is rated at 15 kWp, with a battery energy storage system consisting of 24 2V 1236 Ah lead-acid batteries. It suffers a critical problem of overloading due to the PV generation-loading imbalances. Load-generation imbalances are caused by a mismatch between the actual power generated and power demand. The load-generation disparity caused regular power outages, resulting in an unreliable power supply in the village.

Interactions with the community revealed that, initially in the village, there were no individual customer connection meters, hence they opted for uniform monthly bills for all the microgrid customers, irrespective of their actual consumption and the number and type of their loads. The lack of meters and of access to energy consumption data poses a sustainability challenge for microgrid customers and operators, as customers cannot track their energy consumption and the developers cannot gain any insight into customer demand or provide a customer-side means of limiting power draw to ensure microgrid equipment operational safety and endurance; this similar challenge was raised by the microgrid users. Smart meters, complemented by appropriate digital tools and reliable communications networks and technologies, are thus recommended, to facilitate energy usage monitoring and control, both at metering connection level and appliance level (as discussed by Waswa et al., 2025). This will also support deployment and awareness of energy efficiency measures and energy efficient appliances in microgrids.

The microgrid power quality measurements with a power quality analyser revealed that one of the three phases at the microgrid plant was loaded significantly more than the other two phases, resulting in poorer reliability and more blackouts on the overloaded phase. This highlighted the criticality of effectively training technical personnel and equipping them to access, interpret, and utilize microgrid system data as well as carry out required on-going operation, maintenance, and activities such as line balancing so as to maximize system reliability and minimize avoidable outages. This in turn, contributes to enhanced microgrid sustainability due to better customer satisfaction and retention, as well as increased income from met demand. Incorporation of demand-side management measures is also recommended to better match demand and supply and alleviate microgrid stress during peak load hours.

2.2 Economic Aspects

The community microgrid case study provides a compelling example of how the economic sustainability of rural energy systems is deeply entangled with governance structures, regulatory oversight, and socio-cultural dynamics. Though technically functional at its inception, findings showed that the microgrid now faces declining performance due to financial instability, weakened institutional accountability, and persistent power asymmetries—factors that reflect broader patterns of systemic fragility in decentralized energy development.

One of the core economic challenges lies in the absence of consistent regulatory oversight. Despite being formally under the purview of the District Energy Office and the Rural Energy Agency, these institutions have played a marginal role in enforcing financial discipline, auditing system performance, or supporting community-level governance. It was seen that this regulatory vacuum has allowed localized, informal governance actors—such as technicians, local brokers, and traditional leaders—to assume de facto control over key functions, including collections, disconnections, and conflict resolution. Interactions with respondents revealed that while these actors have sustained day-to-day operations, their legitimacy is informal, their accountability unclear, and their actions often reinforce existing power imbalances.

These dynamics must be understood within the framework of microgrids as socio-technical systems. The microgrid in this case does not operate as a stand-alone technical asset, but as a system embedded in community norms, cultural expectations, and uneven access to knowledge. Most end-users lacked the technical capacity or institutional awareness to question decisions or monitor system performance. This technical dependency reinforced hierarchical governance, where the technician—a single actor—holds disproportionate influenced over both access and pricing, with little oversight from the community or regulators.

Ownership and institutional design compound these issues. Although the microgrid is formally community-owned through the Village Electricity Committee (VEC), the absence of clear handover documentation, limited training, and non-transparent revenue management practices have weakened local ownership. As touched upon by the engineering insight, tariff defaults are widespread. This was stated to be due to not only affordability constraints but also the perception that the system is a donor-funded asset—leading to a »gift mentality« that erodes financial accountability. Without a robust financial model or reserve fund, the system deteriorated technically, with battery storage and panel performance falling below viable levels.

In sum, this Tanzanian case illustrates that the economic sustainability of off-grid energy systems cannot be reduced to revenue sufficiency or hardware reliability. It is shaped by the interplay of regulatory absence, contested ownership, technical dependence, and informal governance structures. Addressing these systemic vulnerabilities requires a deeper institutional and socio-cultural understanding of how microgrids operate on the ground—and a shift in policy focus from infrastructure deployment to governance reform and financial accountability mechanisms that reflect the complexity of community energy systems.

2.3 Social Scientific Aspects

Social issues are commonplace in rural SSA and unpredictable factors with the potential to destabilize low-carbon energy solutions have been recognized by Nuru et al. (2021). The authors further note that while some may be overt, there may also be nuanced social problems requiring careful attention (Nuru et al., 2021, p. 8). Community acceptance plays an important role in the project's successful implementation and has a significant influence on the sustainability of these renewable energy systems. Social acceptance arises from the level of satisfaction within communities. However, it may be hindered by the lack of access to information, which contributes to the lack of awareness (Come Zebra et al., 2021, p. 16). Community participation should be initiated during the system's inception (e.g., by the initial site visit) to build trust and gain context for the project—both key drivers of a project's future success. Explaining the benefits of the project to the community and agreeing on community contributions can yield greater interest and investment. Early and regular meetings and discussion forums between developers, locally respected figures, and community members early in the project will increase the sense of community buy-in and lower the project costs of labour, maintenance, monitoring, and rule enforcement. These meetings cannot be only with village leaders as this increases the risk of elite capture. Developers should use community input when assessing the needs and values of the customers (Gill-Wiehl et al., 2022, pp. 10–11). More so, when gender justice is not ensured during the early stages of project planning, the resulting energy solutions often fail to meet most of women's pressing concerns, resulting in underutilization of the provided electricity (Karlsson, 2019).

The microgrid in this case study was installed in a rural community primarily for lighting, phone charging, and powering productive energy enterprises, but women in the area continued to rely on traditional biomass for cooking and lighting. Such scenarios arise due to the high cost of electrical appliances and a lack of affordable, alternative cooking solutions within the project package (Winther et al., 2020). In many East African households, women are responsible for a wide range of energy-related tasks, including cooking, heating, lighting, and water fetching (Clancy, 2016). Despite their central role in managing household energy, women are often excluded from the decision-making processes surrounding mini-grid projects. For example, a project designed to introduce solar energy for household lighting may not take into account women's preference for cleaner cooking energy, which is often a greater need. Establishing a clear local governance structure or adapting an existing one (e.g., committees with bylaws, specific meeting times, etc.) that caters to the community being served provides an accountability model for successful operation. Project failure is more likely to occur in the Operation and Maintenance (O&M) phase, and thus community participation could keep all stakeholders accountable (Gill-Wiehl et al., 2022, p. 10).

In the case study of the microgrid community in Tanzania, the interviews and focus group discussions revealed that the project was initiated by the national government and introduced to the local community by the Ministry of Energy and district leadership officials. The project's beneficiaries were already decided by the time of the microgrid's introduction, both in the number of 60 beneficiaries and based on their proximity to the selected beneficiary social institutions. Baseline and community needs assessments and project inception meetings were not conducted. A rules and guidelines book compiled by the national government and Ministry of Energy served to educate the stakeholders about their various roles, obligations, and procedures of microgrid operation.

> Not all of the first selected sixty project beneficiaries accepted the project, others refused to be connected to the electricity because they feared it and they didn't have knowledge, when they see the white people they say that, »I did not ask for this electricity, I don't need it.« So, they did not get the electricity (…).
>
> (…) some of them they know that this electricity, they are getting for free because of the political campaign, so when you go there to take the money, they are say[ing], »No, this is electricity that is for free and we got it as a promise of the campaign, we get it for free...« So it was a big challenge to the leaders… they were not willing to pay money for it.

This situation, as expressed during an interview with a local community leader, revealed some of the multiple complexities that the microgrid community faced. It was stated that at the project's inception, many people hesitated to enrol, because they did not understand it; moreover, they had some knowledge and misconceptions about it, such as that the project required human blood sacrifices to succeed, and so they would die if they joined it. Despite the existence of the guidelines book, the rules were not followed at the operation stage; as seen in the earlier indicated engineering and economic aspects, many customers refused to pay for the electricity after consumption. The electricity committee members who were in charge of the physical collection of post-consumption electricity fees shared that they were constantly met with verbal, physical, and spiritual (witchcraft) violence from the electricity fee defaulters. At times, some of the electricity committee members would not keep proper records and therefore would demand money from customers who had already paid. The situation in the community was stated to come to a point that community members were no longer willing to volunteer to serve as electricity committee members and »...only about forty or so households remained with electricity, others had been disconnected from the electricity, even the few households that remained on the grid were not consistent in remitting their utility bills (...)«, as a community leader recalled. The Ministry of Energy and district administration officials that were tasked with the supervisory roles did not carry out these tasks and did not offer the recommended guidance and oversight.

The community leaders and electricity users did not participate in compiling the microgrid operation guidelines. Also, the new customers who were added to the grid after the defaulters and/or deserters had been disconnected did not receive copies of the rules book. Coupled with the fact that the majority of community members did not participate in the initiation stage of the project, there was a lot of operation ambiguity in the community. Every stakeholder seemed to be a law unto themselves. Some electricity users would refuse to pay utility bills; electricity committee leaders and local leaders diverted project resources to other stated tasks and loans for community development—which were never paid; the technician appeared in the community at will and was not answerable to the community leadership; and community members were unaware of their mandate to question and add their voice to project decisions. The spillovers from these stated social aspects did not just reflect existent stakeholder conflicts and hinder access to electricity services; they could also be traced in the economic, technical, knowledge, and expertise struggles of this community-managed microgrid project.

Another hindrance to long-term sustainability is the lack of gender integration during the planning, implementation, and operation of these projects. This is evident in the failure to incorporate the views and needs of women (just as those of men and others), who play crucial roles in energy consumption, household management, and community leadership, which creates misaligned project objectives and low community engagement, ultimately undermining the success of these initiatives. The economic empowerment of women is another critical factor in the success of mini-grid projects. Indeed, energy access can serve as a catalyst for economic development, enabling people to engage in entrepreneurial activities, such as running small businesses or using energy-powered equipment for agricultural processing. However, when women are excluded from the project design and implementation phases, they may miss out on the economic benefits that stem from energy access. For instance, in Tanzania, the microgrid project under review here that sought to provide energy for small-scale industries did not consider the specific needs of women entrepreneurs. As a result, while men were able to use the energy (at least initially, before the microgrid weakened) in their enterprises like saloons, restaurants, cinemas, and similar enterprises, women lacked access to the energy resources necessary for their businesses, such as food processing or textile production. This disparity hinders the ability of women to benefit from the project economically, further entrenching gender inequalities (Clancy & Mohlakoana, 2020).

In addition to economic impacts, gender integration is essential for fostering social acceptance and ensuring the overall success of microgrid projects. Women, as the primary users of household energy, are critical to ensuring that new energy systems are adopted and used effectively. The microgrid in the case study failed to engage women early on in the planning process and faced initial resistance from the community. This resistance partly stemmed from a disconnect between the energy services provided and the actual needs of households. For example, it was met with hesitation when women, who were the main consumers of household energy, felt that the energy provided did not address their daily tasks, such as cooking and food preservation. A similar situation happened in Uganda where, after microgrids were set up, instead of adopting the new systems, many households reverted to traditional energy sources like firewood and kerosene, leading to poor uptake of the microgrid services (Njoh & Akiwumi, 2011). This highlights the importance of involving women in all stages of the project cycle to ensure community buy-in and proper utilization of the energy services offered.

Gender integration in mini-grid projects is not without its challenges. Cultural norms and discourses in many East African societies often restrict women's participation in public decision-making, relegating them to traditional roles that limit their visibility in energy planning and policy processes (Elnakat & Gomez, 2016). Additionally, women face barriers to education and technical training, which further excludes them from meaningful involvement in energy initiatives. A study conducted in East Africa for this publication found that even though women were primarily responsible for managing household energy, they were rarely included in community meetings or discussions about the installation and operation of microgrid systems. Overcoming these barriers requires deliberate efforts by project developers and policymakers to include women in capacity-building initiatives, provide them with leadership opportunities, and ensure that energy policies are gender-sensitive. For example, training women as technicians or operators of microgrid systems can help to break down gender barriers and promote their involvement in the energy sector (Tsagkari, 2022).

Indeed, successful examples of gender equity in energy projects demonstrate the transformative potential of such approaches. For instance, in Zanzibar under the auspices of the barefoot college, microgrid projects supported by the government and partners prioritized the inclusion of women at every stage of the project (Mininni, 2020). Women were not only involved in the planning process but also trained as technicians and operators of the microgrid systems. This active participation has led to higher community engagement, improved household energy access, and increased economic opportunities for women, such as the ability to start small businesses that utilize the new energy services (Winther et al., 2020). This case illustrates that when women are meaningfully involved, microgrid projects are more likely to succeed and provide broader development benefits (Mininni, 2022).

2.4 Didactics of Technology Aspects

Aspects related to the didactics of technology (DT), such as sensitization and training/skilling, can contribute to energy development sustainability. Scholars such as Bayram (2018), Dauenhauer & Frame (2017), and Namaganda-Kiyimba & Mutale (2018) propose factors that impact the sustainability of energy development projects such as microgrids. The factors include technological, financial, social, organizational, environmental, legal, regulatory, health and safety, ethical, physical, and political. The factors resonate with the PESTEL framework by Come Zebra et al. (2021). The DT aspects majorly contribute to the social and technological factors through knowledge transfer. Knowledge transfer contributes to local capacity-building that ensures: (1) Availability of a pool of future energy engineers with the expertise to design, install, operate, and maintain energy system networks; (2) Sensitization and skilling aimed at triggering a mindset change so that community inhabitants have divergent views regarding benefiting from electrical energy. For example, instead of limiting electrical energy application to the provision of light at night, charging phones, and powering radios, inhabitants can be trained to create solutions and businesses tailored to community challenges. In essence, promoting productive use of energy (PUE) while ensuring effective and efficient electrical energy utilization results in improved demand, monthly consumption, and income levels since consumers earn money while utilizing electrical energy. As Booth et al. (2018) confirm, PUE is a key success factor for energy development projects.

Therefore, knowledge transfer is a vital element in the sustainability of energy projects. Indeed, researchers such as Boche et al. (2022, pp. 4, 17–18) confirm that microgrids-related knowledge transfer is a sustainability element. To contribute to knowledge transfer, DT aspects include the development of social-technical strategies to build local technical human capacity in EA for sustainable energy development. One such social-technical strategies is the development of locally relevant and community-friendly training programmes (cf. Huh, 2017, p. 1). Such programmes are those that (1) aim to solve challenges faced by the locals, (2) embrace the training interests and needs of the community inhabitants, and (3) consider the skills of the local inhabitants among others. Accommodating these considerations, the training programmes are highly likely to receive acceptance and be trusted by the locals.

In light of the foregoing discussion, and considering the specific case study, the development of the training programme necessitates establishing the local inhabitants' electrical energy utilization challenges, training interests, transferrable skills gained from previous trainings, if any, and their training needs. Data was collected via a questionnaire from 39 consumers out of the 60 connected to the community-owned microgrid. Considering electrical energy utilization challenges, analysed responses reveal that consumers in the village face the following challenges: Power outages (70 %), a non-transparent payment system (9 %), poor power quality (7 %), business flow interruptions (6 %), high electricity cost (4 %), poor installation workmanship (2 %), and technical Human Resource (HR) related challenges (2 %).

Challenges related to power outages recorded the highest frequency (70 %). It should be noted that the microgrid under study is a 2015 commissioned facility and thus has deteriorated battery storage and visibly damaged solar panels, which contribute to the blackouts. Power outage challenges and poor power quality can be better addressed by operators and microgrid owners from the generation and distribution subsystems. According to Bogere (2024, p. 67),

> Knowledge and skills aimed at minimizing blackouts, for example ensuring load balancing ought to be imparted to BSc Elec Eng students and technicians. To minimise outages at a consumer's premises, consumers can be skilled to carry out simple maintenance tasks and ensure the safety of selected components.

In addition, for consumers to contribute to the minimization of power outages, they ought to be sensitized about, and trained to avoid, illegal connections to the power network, and shown that efficient electrical energy utilization and the use of only acceptable electrical loads can minimize power outages. Challenges related to business flow interruptions are a result of power outages and poor power quality. Addressing power outage and poor power quality intuitively improves business operations and thus addresses power flow interruption challenges.

The high cost of electricity is another challenge mentioned by consumers. Although a monthly flat charge of Tanzanian Shillings 3,000 (approx. €1.14) is levied from each consumer, 4 % of the consumer responses reveal that electricity cost is high. However, as evident in Fig. 1, lights were observed in the community during daytime. The usual practice of not switching off lights during daytime and the use of high-rating electric bulbs add to the high electricity costs.

Fig. 1: Lights kept on during daytime in the Tanzanian village—many were observed during data collection (reproduced from Bogere, 2024, p. 67).

Although consumers are charged a flat fee, some complain about the high electricity cost. Some of the consumers who complain say that it is unfair to charge commercial and residential consumers a flat fee when the consumption is not metered and thus not considered. Residential consumers struggle to pay the monthly bill. This presents a didactical opportunity to sensitize and skill consumers on how to strategize and generate income from electrical energy consumption. Therefore, consumers need knowledge and skills in PUE, and entrepreneurship, among others.

The previously discussed challenges and others, such as poor installation workmanship and technical HR related challenges, can be minimized if consumers are appropriately trained. An example of poor workmanship is evident in Fig. 2 where circuits for lights and socket outlets are terminated within the same junction box (JB) and use the same cable sizes, that is 1.5 mm^2. Best electrical installation practices recommend separating light and socket electric circuits. As Bogere (2024, p. 68) argues, if consumers are equipped with basic installation skills, they can ensure that technicians carry out electrical installations using appropriate cable sizes. Relatedly, consumer statements like »we do not have a technician in the village«, »expensive cost needed to pay transport for technicians to come (...)«, »technicians are far, recruit more workers to support« speak to the need for training. This necessity becomes even more telling as some consumers revealed their interest to be trained so as to work as electricians. Lessons from the Barefoot College International, where illiterate women are trained to take on such responsibilities in rural communities, attest to the fact that it is possible for trained local inhabitants to solve technical HR-related challenges. Therefore, equipping consumers with electrical installation skills would greatly reduce technical HR-related challenges.

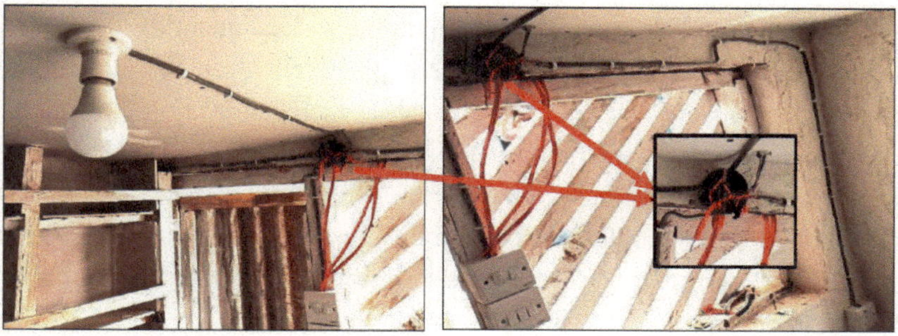

Fig. 2: Light and socket circuits terminated in the same JB and using the same cable size—1.5 mm^2 (Bogere, 2024, p. 69).

To this end, electrical energy consumers need to be equipped with appropriate skill sets to minimize electrical energy utilization challenges and support sustainability in energy development. Similarly, responses to open-ended questions were analysed to establish knowledge themes and thus skill sets to meet consumers' training needs, training interests, transferable skills, and the required skills for repair and maintenance services. The skill sets enable the development of appropriate knowledge content to guide the skilling of consumers to acquire the relevant competencies. According to Bogere (2024, p. 73), acquiring the relevant competencies entails training consumers to: adapt efficient and cost-effective electrical energy utilization approaches; perform simple electrical installations for microgrids sustainability in communities; adhere to electrical health and safety rules in electrical-related undertakings while being mindful of environmental concerns; acquire entrepreneurship knowledge and skills for income generation while consuming electrical energy and solving local problems. Relevant competencies therefore are expressed as the training objectives the development of which is guided by the revised Bloom's taxonomy (Anderson et al., 2001). The training objectives characterized by the intended learning objectives (ILOs) form the skill sets envisaged to enable electrical energy consumers to acquire the relevant competencies to support sustainability in energy development. The skill sets and the ILOs then inform the development of the knowledge content and/or training programme.

In summary, DT aspects that contribute to sustainability in energy development include the establishment of electrical energy utilization challenges, the development of knowledge content, and training to enable consumers to acquire the relevant competencies. Training and acquisition of competencies by consumers contributes to SDG target 4.7, the aim of which is to contribute to sustainability in energy development. With sustainable energy, energy access levels improve and thereby contribute to SDG 7.

2.5 Conclusion

The presenting aspects and the ensuing discussions surrounding the Tanzanian case study here highlight that sustainable implementation of a microgrid cannot be reduced to a single issue, such as revenue sufficiency, hardware reliability, community participation, or technical awareness and skill. Rather, due attention needs to be paid to aspects relating to the technical, the economic, the social, and didactics of technology. Failing to adequately account for any of these key aspects is likely to result in continually emerging conflicts and implementation discrepancies, which will jeopardize the efficient operation and general sustainability of a microgrid project.

2.6 References

Agoundedemba, M., Kim, C. K., & Kim, H. G. (2023). Energy status in Africa: Challenges, progress and sustainable pathways. *Energies, 16*(23). https://doi.org/10.3390/en16237708

Anderson, L. W., Krathwohl, D. R., Airasian, P. W., Cruikshank, K. A., Mayer, R. E., Pintrich, P. R., Raths, J., & Wittrock, M. C. (2001). *A Taxonomy for Learning, Teaching, and Assessing: A Revision of Bloom's Taxonomy of Educational Objectives*. Addison Wesley Longman, Inc.

Bayram, I. S. (2018). Teaching smart power grids: A sustainability perspective. *12th International Conference on Compatibility, Power Electronics and Power Engineering (CPE-POWERENG)*.

Boche, A., Foucher, C., & Villa, L. F. L. (2022). Understanding microgrid sustainability: A systemic and comprehensive review. *Energies, 15*(8), 1–29. https://doi.org/10.3390/en15082906

Bogere, P. (2024). *Knowledge Transfer for Microgrids Sustainability: The Case of East Africa* [Doctorate, Paderborn University]. https://doi.org/10.17619/UNIPB/1-2167

Booth, S., Li, X., Baring-Gould, I., Kollanyi, D., Bharadwaj, A., & Weston, P. (2018). *Productive Use of Energy in African Micro-grids: Technical and Business Considerations*. www.nrel.gov/publications.

Clancy, J. S. (2016). Gender and household energy: The international context. *Energy Research & Social Science, 18*, 35–44. https://doi.org/10.1016/j.erss.2016.02.004

Clancy, J. S., & Mohlakoana, N. (2020). Gender audits: An approach to engendering energy policy in Nepal, Kenya and Senegal. *Energy Research & Social Science, 62*, 1–19. https://doi.org/10.1016/J.ERSS.2019.101378

Come Zebra, E. I., van der Windt, H. J., Nhumaio, G., & Faaij, A. P. C. (2021). A review of hybrid renewable energy systems in mini-grids for off-grid electrification in developing countries. *Renewable and Sustainable Energy Reviews, 144*(March). https://doi.org/10.1016/j.rser.2021.111036

Dauenhauer, P. M., & Frame, D. F. (2017). Sustainability analysis off-grid community solar PV projects in Malawi. *GHTC 2016—IEEE Global Humanitarian Technology Conference: Technology for the Benefit of Humanity, Conference Proceedings*, 113–120. https://doi.org/10.1109/GHTC.2016.7857268

Elnakat, A., & Gomez, J. D. (2016). Energy engenderment: An industrialized perspective assessing the importance of engaging women in residential energy consumption management. *Energy Policy, 92*, 133–144. https://doi.org/10.1016/j.enpol.2016.01.046

Gill-Wiehl, A., Miles, S., Wu, J., & Kammen, D. M. (2022). Beyond customer acquisition: A comprehensive review of community participation in mini grid projects. Renewable and Sustainable Energy Reviews, 153(October 2021). https://doi.org/10.1016/j.rser.2021.111778

Huh, T. (2017). Analyzing the configuration of knowledge transfer of the green island projects in S. Korea. *The Korean Journal of Policy Studies, 32*(1), 1–26. https://doi.org/10.52372/kjps32101

IEA, IRENA, UNSD, World Bank, & WHO. (2024). *Tracking SDG7: The Energy Progress Report 2024*. IEA. https://www.iea.org/reports/tracking-sdg7-the-energy-progress-report-2024

International Energy Agency. (2023). *Africa Energy Outlook 2022: World Energy Outlook Special Report (Revised in 2023)*. 250. https://www.iea.org/reports/africa-energy-outlook-2022%0Ahttps://iea.blob.core.windows.net/assets/220b2862-33a6-47bd-81e9-00e586f4d384/AfricaEnergyOutlook2022.pdf

Karlsson, G. (2019). Engendering the energy transition: Climate justice and women's rights in renewable energy projects. *Energy Research & Social Science, 60*, 101341. https://doi.org/10.1016/j.erss.2019.101341

Krauter, S., Rohrer, N., Lehmann, S., Klaus, T., Böcker, J., Freitag, C., Hehenkamp, B., Hilleringmann, U., & Temmen, K. (2019). *Africa: Research and Teaching Platform for Development—Sustainable Modular Grids for Grid Stability (ART-D Grids).*

Mininni, G. M. (2020). *Can Energy Empower Women? A Case Study of the Solar Mamas' Program in Rural Rajasthan, India*. Keele University. https://keele-repository.worktribe.com/output/418112

Mininni, G. M. (2022). The Barefoot College »eco-village« approach to women's entrepreneurship in energy. *Environmental Innovation and Societal Transitions, 42*, 112–123. https://doi.org/10.1016/J.EIST.2021.12.002

Namaganda-Kiyimba, J., & Mutale, J. (2018). Sustainability metrics for rural electrification in developing countries. *IEEE PES/IAS PowerAfrica Conference*, 426–431. https://doi.org/10.1109/PowerAfrica.2018.8521140

Njoh, A. J., & Akiwumi, F. A. (2011). Colonial legacies, land policies and the millennium development goals: Lessons from Cameroon and Sierra Leone. *Habitat International, 35*(2), 207–218. https://doi.org/10.1016/j.habitatint.2010.09.007

Nuru, J. T., Rhoades, J. L., & Gruber, J. S. (2021). The socio-technical barriers and strategies for overcoming the barriers to deploying solar mini-grids in rural islands: Evidence from Ghana. *Technology in Society, 65*(August 2020), 101586. https://doi.org/10.1016/j.techsoc.2021.101586

Tsagkari, M. (2022). The need for gender-based approach in the assessment of local energy projects. *Energy for Sustainable Development, 68*, 40–49. https://doi.org/10.1016/J.ESD.2022.03.001

Waswa, A. M., Namaganda-kiyimba, J., Mabala, S. J., Seremba, N. E., Ssemakula, F., Mark, G., Kakande, J. N., & Ndagire, E. (2025). Development of a smart meter for deployment in on-grid and off-grid energy systems. *2025 IEEE International Conference on Consumer Electronics (ICCE)*; IEEE, 1–6. https://doi: 10.1109/ICCE63647.2025.10930148

Winther, T., Ulsrud, K., & Saini, A. (2020). Solar powered electricity access: Implications for women's empowerment in rural Kenya. *Energy Research & Social Science, 60*, 101338. https://doi.org/10.1016/j.erss.2019.101338

3. Improving Sustainable Energy Engineering

3.1 Using and Storing Sustainable Energy

Josephine Nakato Kakande, Godiana Philipo, Tobias Klaus

3.1.1 Introduction

As the world grapples with the twin crises of energy scarcity and climate change, Africa stands at a critical juncture. The global average surface temperature has increased by about 1.5°C (2.7°F) above pre-industrial levels, defined as the 1850–1900 average (WMO, 2025).

Major cities in East Africa have already experienced significant warming since pre-industrial times, with temperature increases nearly double the global average. For example, data from ICPAC (2021) show the following temperature changes since 1860:

- Addis Ababa (Ethiopia): +2.2°C
- Dar es Salaam (Tanzania): +1.9°C
- Mogadishu (Somalia): +1.9°C
- Nairobi (Kenya): +1.9°C.

Climate models project that by the end of the 21st century, if emissions continue at current rates, East Africa could see temperature rises in the range of 2°C to 6°C above pre-industrial levels, depending on the emissions scenario.

Specifically for Tanzania, projections indicate that by the late 21st century (2090s), average temperatures during June to August could be 2.2°C to 2.6°C higher than in the late 20th century (Magang et al., 2024).

The median temperature is predicted to increase 1.5 times the global mean by 2100 (Ddamulira, 2017). Although Africa makes an insignificant contribution to global warming (less than 3 % of the world's total greenhouse emissions), it remains the most vulnerable continent to adapt to climate change variability due to several factors such as limited technology and infrastructures, endemic poverty, etc. (Parry & Palutikof et al., 2007).

Fig. 3: The African map showing the three national states of Tanzania, Uganda, and Kenya, which were the focus of this study. Source: Danny (2004).

The effects of climate change, such as rising temperatures, shifting precipitation patterns, and increased extreme weather events, pose challenges to East Africa's energy security (Xie & Xie, 2024). East Africa is a region characterized by diverse ecosystems, growing populations, and burgeoning energy demands.

East Africa currently comprises seven countries. Tanzania, Kenya, and Uganda were the traditional members back in 2000, followed by Rwanda and Burundi, who joined in 2007; South Sudan in 2016; and in 2023, the Democratic Republic of Congo joined as the seventh country (KFW Development Bank, n.d.). This chapter makes the explicit choice of referring to EA with the three traditional states of Tanzania, Uganda, and Kenya (Fig. 3). With its abundant renewable resources and growing energy needs, the region has the potential to become a global leader in sustainable energy. This chapter explores the promising landscape of sustainable energy use and storage in East Africa, highlighting the region's unique opportunities associated with using sustainable renewable energy.

3.1.2 Context and Background

East Africa is a region of diverse landscapes and rich natural resources. The climate ranges from tropical to temperate in the highlands of Tanzania, tropical to arid with two rainy and dry seasons in Kenya, and tropical to semi-arid in the North-Eastern part of Uganda (Othieno & Awange, 2016, Ch. 2). The geographical diversity presents both challenges and opportunities for energy generation, storage, and use. While Uganda is a landlocked country, both Tanzania and Kenya have the coastline of the Indian Ocean. Lake Victoria is the world's second-largest freshwater lake, the common border of the three countries. East Africa is gifted with several energy resources whose potential has not been fully utilized/exploited. The presence of various rivers gives potential for hydropower generation, while geothermal and oil deposit sites, coal, and natural gas have been identified, with some exploitation already underway. For instance, Tanzania and Kenya's Rift Valley are home to significant geothermal resources, while Tanzania's coastline offers substantial wind energy prospects (Othieno & Awange, 2016).

Despite the fact that the EA region is endowed with vast renewable energy potential, it still relies heavily on traditional biomass for energy, posing significant environmental and health risks. It is reported that non-clean energy sources contribute over 80 % of the total energy consumption in East Africa (Othieno & Awange, 2016). Sustainable energy can drive economic growth by creating jobs, attracting investment, and reducing energy costs. For example, the Lake Turkana Wind Power project in Kenya has not only provided clean energy but also created employment opportunities for the local community. The shift to renewable energy sources can significantly reduce greenhouse gas emissions, mitigating the effects of climate change. Furthermore, sustainable energy reduces reliance on deforestation for fuel, preserving East Africa's rich biodiversity.

3.1.3 Renewable Energy Sources in East Africa

3.1.3.1 Hydropower

Hydropower remains a crucial energy source, which acts as the primary source of electricity in the East African region. Hydropower can be adapted quickly to changes in demand and counteract fluctuations in power generation, such as in wind or solar power. Many hydropower plants feature a water reservoir, so they can also serve as an energy storage unit. Tanzania has excellent potential for hydropower generation due to precipitous river availability due to its inland being at a higher elevation than the sea level. Most of the potential hydropower sites are the rift valley escarpments, rivers, and basins across the country, as well as lakes such as Lake Victoria, Tanganyika, Nyasa, Rukwa, Manyara, Eyasi, and Natroni (Kichonge, 2018). The total approximated hydropower potential is 38,000 MW, which is equivalent to 190,000 GWh/year (Kichonge, 2018). Some of the already operational and ongoing projects for hydropower plants in Tanzania are listed in Tab. 1 (Tanzania Power Production and Demand—2024 Update, n.d.).

In Uganda, hydropower comprises 78 % of the installed capacity (Electricity Regulatory Authority (ERA), 2023), while in Kenya hydro comprises only 25 % of the installed capacity (Kenya National Bureau of Statistics, 2024).

Tab. 1: Hydropower plants in Tanzania (operational, planned, and ongoing)

Name	Capacity	Type	Location	Status
The Kidatu Plant	204 MW	Reservoir	Great Ruaha river/basin in Morogoro	Opera-tional
The Kihansi Plant	180 MW	Reservoir	Kilombero river, Morogoro	Opera-tional
Mtera plant	80 MW	Reservoir	Great Ruaha river/basin in Dodoma	Opera-tional
New Pangani falls	68 MW	Reservoir	Pangani River, Tanga	Opera-tional
Hale	21 MW	Run of the river	Pangani river/basin in Kilimanjaro	Opera-tional
Nyumbaya Mungu	8 MW	Reservoir	Mwanga, Kilimanjaro	Opera-tional
Uwemba	0.84 MW	Run of the river	Njombe	Opera-tional
The Julius Nyerere Hydropower Station (JNHPP)	2115 MW	Reservoir	Stiegler's Gorge, Moro-goro Region (Rufiji River)	Partially opera-tional
Malagarasi plant	49.5 MW	Run of the river	Malagarasi river, Kigoma	Planned
Kakono plant	87.8 MW	Run of the river	Kagera River, Kagera	Planned
Rusumo plant	80 MW	Run of the river	Kagera River, Kagera	Opera-tional

Tanzania's hydropower potential is considerable but varies with water flow. According to the University of Dar es Salaam, the country's two largest reservoirs, *Mtera* and *Nyumba ya Mungu*, have storage capacities of 3,200 million m^3 and 580 million m^3, respectively. Tanzania's hydropower potential is estimated at 4.7 GW, with the Julius Nyerere hydropower project, which is currently under construction, aiming to add 2,115 MW to the grid.

As of the end of 2024, Uganda's installed hydro capacity was 1,714 MW with large hydro comprising 1,500.2 MW and mini-hydro 241.3 MW (Electricity Regulatory Authority (ERA), 2025). For Kenya, at the end of 2024, the country's installed hydro capacity was 872.5 MW (EPRA, 2025). In East Africa, there are no pumped storage hydropower plants.

3.1.3.2 Solar Power

East Africa's high solar isolation makes it an ideal location for solar power generation. The average daily solar irradiance is 5.5 kWh/m^2. A significant advantage is the low variation between a typical day in the worst month of the year (4–4.5 kWh/m^{2}d) during the rainy season, and a typical day of the best month of the year (6–7 kWh/m^{2}d). This eliminates the need for seasonal energy storage, which is necessary, e.g., for European locations. Several studies revealed that the Eastern Africa region has an annual estimate of 175 PWh and 220 PWh for Concentrated Solar Power (CSP) and Photovoltaics (PV), respectively, being the highest technical potential for solar power technologies in the region. Overall, the sum of land area of the following countries, taking into account all suitable areas (Burundi, Djibouti, Eritrea, Ethiopia, Kenya, Rwanda, Somalia, Sudan, Tanzania, and Uganda), is 6,225,847 km^2. Among the three selected East African countries, Kenya is generally considered the leader in large-scale solar projects development compared to Tanzania and Uganda. In Tanzania, the annual technical solar power potential is estimated to be 31,482 TWh for CSP technology and 38,804 TWh for PV technology (this is the overall sum, taking into account all suitable areas). The referenced data are from IRENA working paper 2014 (Hermann et al., 2014) Solar resources in Tanzania are more concentrated in the central regions such as Dodoma, Shinyanga, and Singida. Tanzania's progress on solar adaptation is improving through planned and already established private and public owned solar systems. The first successful solar project with 150 MW of electrical power to be connected to the grid will revolutionize the Tanzanian energy mix.

In Uganda, the average solar radiation is estimated to be around 5.1 kWh/(m^{2}day) making it the most adoptable renewable energy source in the country. A study based on measured data concluded that the average global horizontal irradiation is 1,680 kWh per year with variability below 10 % daily, and an annual variation not exceeding 20 %. The low variability and above threshold for solar energy gives higher potential for large-

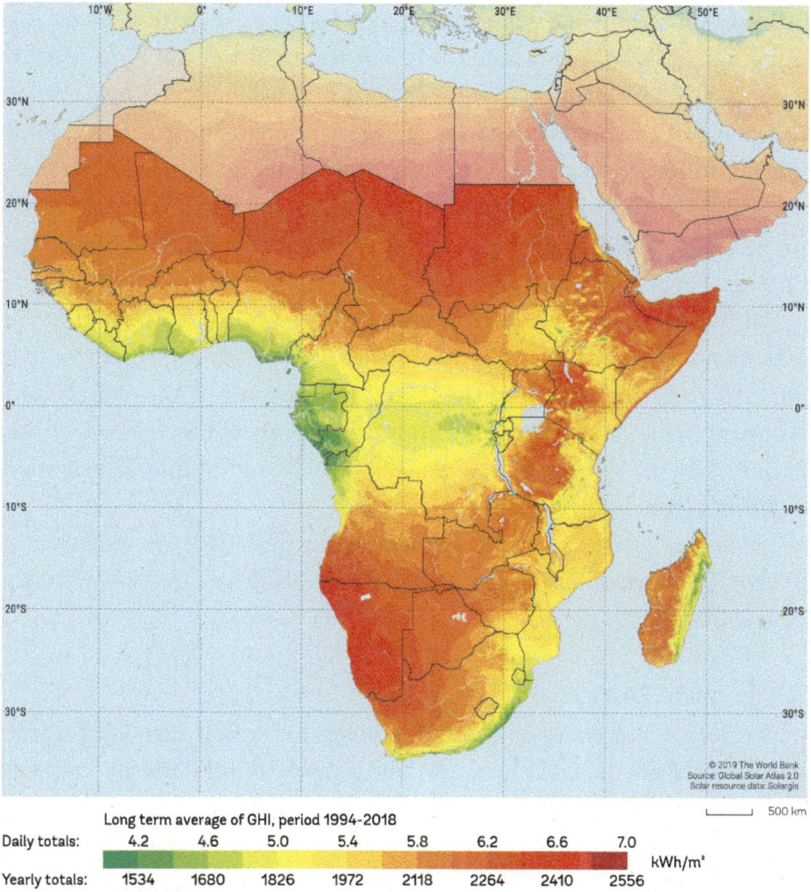

Fig. 4: Solar irradiance values (global irradiance, horizontal) of Africa (World Bank Group / SolarGIS 2019).

scale solar PV plants with minimum backup requirements (Katongole et al., 2023). The majority of the country (69.3 %), which is comprised of the Central, Eastern, and Northern regions, has favorable potential for large-scale solar PV power plants and CSP facilities.

Kenya's estimated total solar power potential is around 15,000 MW with an average daily irradiance of 5 kWh/m^2d making it one of the most promising locations for solar energy development in Africa. According to literature, 34 %, 26.5 %, and 10.1 % of the total Kenyan land receives between 5–5.5 kWh/m^2d, 5.5–6 kWh/m^2d, and 6.0–6.5 kWh/m^2d, respectively. One of the notable Kenyan solar power plants in worth 13 billion Kenyan shillings (Kshs.) (86.4 million Euros), in Garissa County. The plant has 50 MW installed capacity with more than 200,000 PV panels which can generate up to 2 % of the country's energy mix (Kariuki & Sato, 2018; Katongole et al., 2023).

An overview of the solar irradiance values (global horizontal irradiance) for medium daily and annual values is given in Fig. 4. The map is obtained from the Global Solar Atlas 2.0 on behalf of the World Bank Group, utilizing SolarGIS data, with funding provided by the *Energy Sector Management Assistance Program* (ESMAP) (for additional information: see https://globalsolaratlas.info).

While the relevant locations discussed in this book are close to the equator, the horizontal irradiance values are also the values relevant for PV installations.

3.1.3.3 Wind Energy

The region's wind energy potential is being harnessed through projects such as the *Lake Turkana* in Kenya and *Singida Wind Farm* in Tanzania. These initiatives showcase the feasibility and benefits of wind power in East Africa.

Wind has contributed to the growth of installed capacity and available energy to African grids that have capacity and energy constraints. The *Lake Turkana wind farm* contributes 17 % of Kenya's installed wind power capacity. The 2.4 MW *Mwenga Wind Farm* in Tanzania and an isolated grid running primarily on a mini-hydro plant of 4 MW account for wind farms in Tanzania. Wind farms provide the needed capacity and energy to increase the level of electricity supply in the countries looking to expand supply and access to electricity, with the added benefit of supplying renewable power. As such, wind has been particularly effective in firming up solar supply. The farm supports supply during periods of low hydrology, as is the case on a large scale with wind and hydropower complementarity in Latin

American countries like Brazil, Uruguay, and Costa Rica (Global Wind Energy Council (GWEC), 2023). Fig. 5 provides an overview of wind energy potential in the form of average wind speeds of the entire SSA.

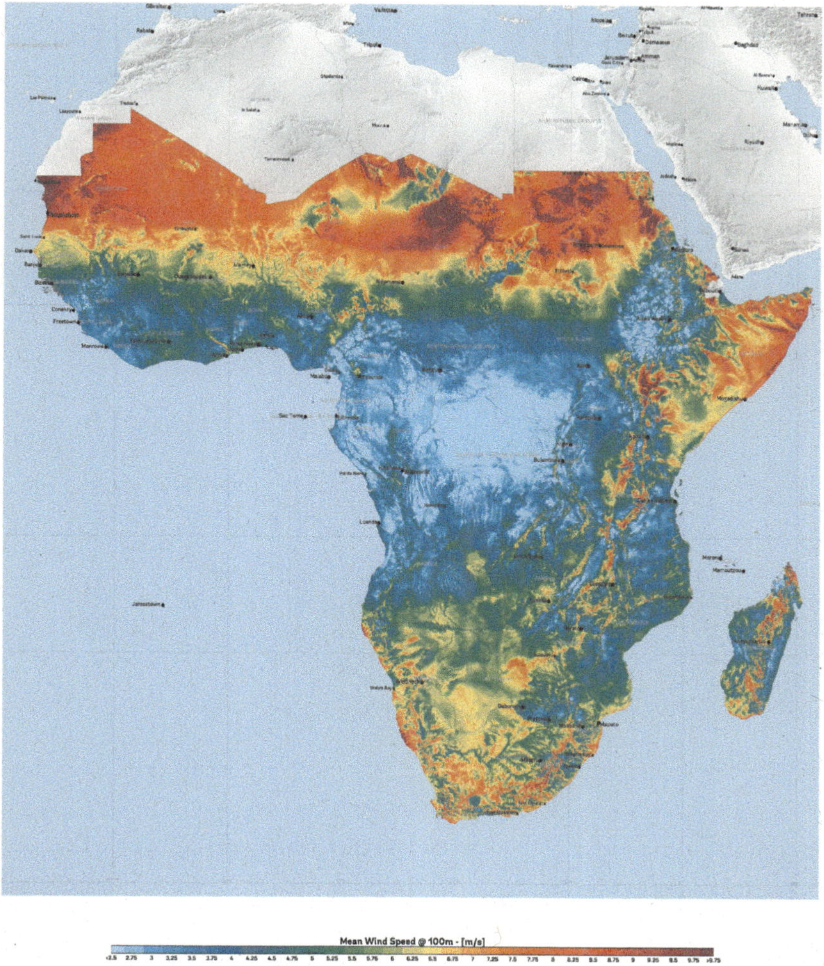

Fig. 5: Wind map of Sub-Saharan Africa: Average wind speeds. (World Bank Group, Global Wind Atlas, DTU 2019).

3.1.3.4 Geothermal Energy

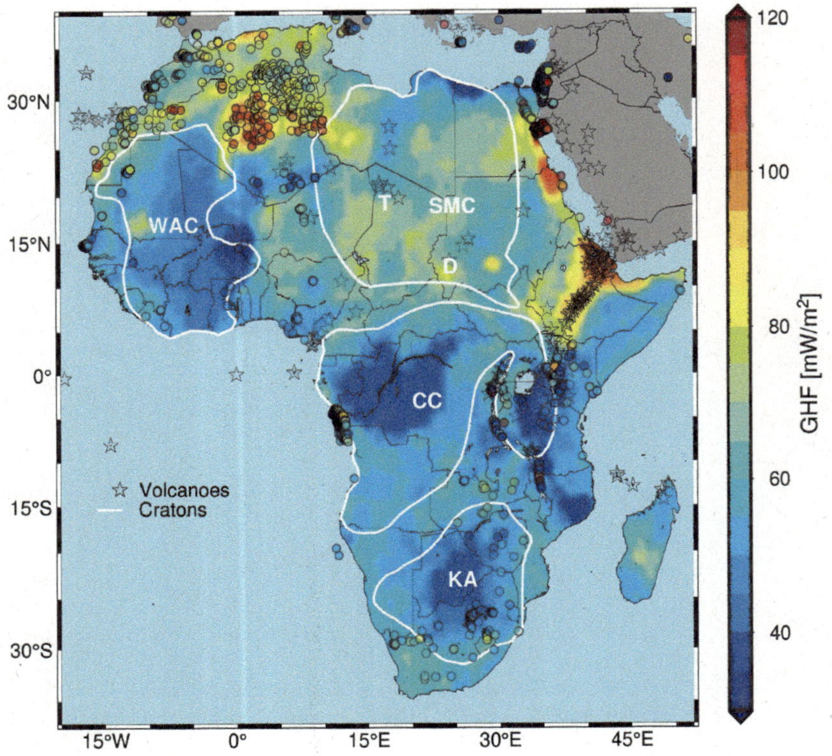

Fig. 6: Geothermal heat flow intensity in Africa. Source: Al-Aghbary et al. (2022).

Kenya leads the way in geothermal energy, with the *Olkaria Geothermal Plant* being one of the most successful projects. Geothermal energy provides a stable and reliable power source, crucial for balancing the intermittent nature of solar and wind energy. The contribution of geothermal energy in Kenya is about 880 MW_e. Initially, Kenya was the second country to install a geothermal power plant with the 45 MW_e Olkaria power plant coming into operation in 1981–1985. The installed capacity in Kenya has grown to 880 MW_e from several sites. Kenya has continued to lead in geothermal development and, in 2017, geothermal represented about

46 % of electricity produced in the country against its installed capacity of about 28 %. Geothermal resource development in Tanzania is at the exploration stage with reconnaissance. Tanzania envisages installing at least 200 MW_e by 2025 from a portfolio of five projects: *Ngozi*, *Kiejo-Mbaka*, *Songwe*, *Luhoi*, and *Lake Natron*. Geothermal resources in Uganda are all located in the Western branch of the East African Rift System (EARS), mainly along the Uganda-DRC border. The main geothermal areas characterized by large hot spring flows are *Katwe-Kikorongo (Katwe)*, *Buranga*, *Kibiro*, *Panyimur*, and *Ihimbo*. Among these, *Kibiro*, *Panyimur*, and *Buranga* are at advanced stages of exploration while the others are at detailed reconnaissance phase. The evaluation of the geothermal systems in Uganda revealed that all the systems are fault/fracture-controlled and low to medium temperatures.

A geothermal heat flow map of Africa is provided in Fig. 6.

3.1.4 East African Energy Mix

Our focus is primarily on Uganda, Kenya, and Tanzania, and their energy mixes are described below.

As of 31 December 2022, the total installed capacity of power in Tanzania was 1,938 MW, 1,899 MW being on-grid installed capacity and 39.3 MW off-grid installed capacity. A report on Tanzania's energy mix showed that Tanzania has diverse power sources including hydro, coal, solar, wind, natural gas, and geothermal, most of which are untapped. In 2023, the component of the grid's installed capacity was mainly covered by natural gas which comprised 1,194 MW, hydropower comprised 602 MW, 84 MW was produced with fuel, and 10.5 MW obtained with biomass. Fig. 7 shows the percentage composition of the total energy mix of the grid installed capacity in Tanzania.

Most of the power is produced by TANESCO, which as of 2022 operated eight natural gas power plants, seven hydropower plants, two heavy fuel oil plants, and seven small gas oil power plants.

In 2022, Uganda had a total electricity generation capacity of 1,402 MW. Renewable energy accounted for 93 % of Uganda's total installed power capacity, with hydropower contributing 78 %, cogeneration from sugarcane bagasse providing 10 %, and solar PV systems making up 5 % (ERA, 2023), as depicted in Fig. 8. *Uganda Electricity Transmission Company Limited* (UETCL) is responsible for all grid transmission while Umeme was the country's biggest distribution utility. The 20-year concession to *Umeme* ended in March 2025, and *Uganda Electricity Distribution Company Limited* (UEDCL) took over the distribution network from *Umeme*.

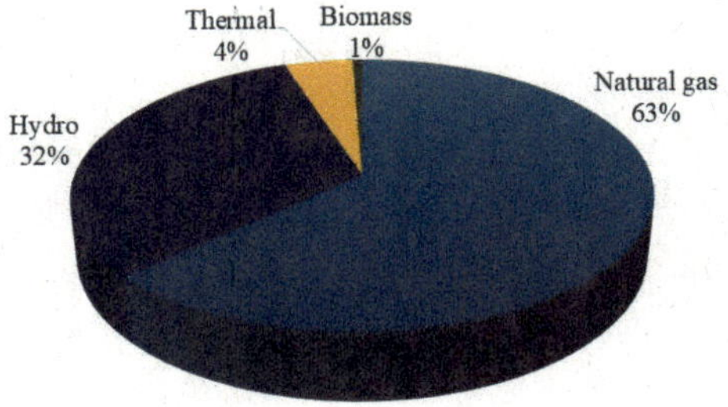

Fig. 7: Percentage composition of the total electricity mix of the grid installed capacity in Tanzania in 2023 (Tanzania Power Production and Demand—2024 Update).

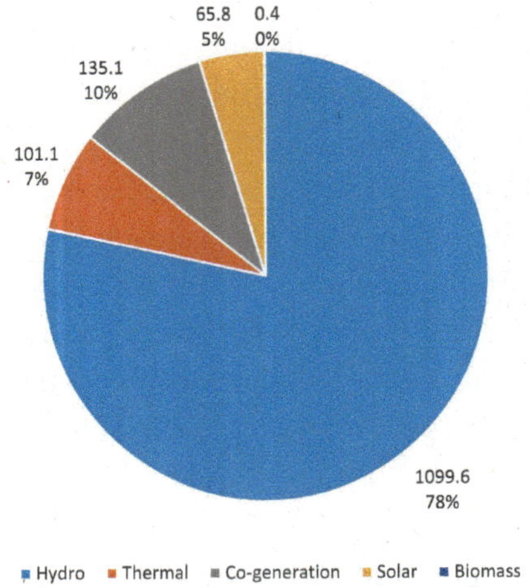

Fig. 8: Uganda's electricity generation capacity (in MW) together with relative share of overall capacity (in %) as of 2022. (ERA, 2023).

In 2023, 6,032 GWh was generated by Uganda's grid-connected plants, of which 93 % was from hydro plants, as illustrated in Fig. 9. In the same year, 35.5 GWh was generated/purchased by Uganda's leading off-grid electricity distributors *West Nile Rural Electrification Company Ltd* (WENRECO) and *Kalangala Infrastructure Services* (KIS) Ltd.

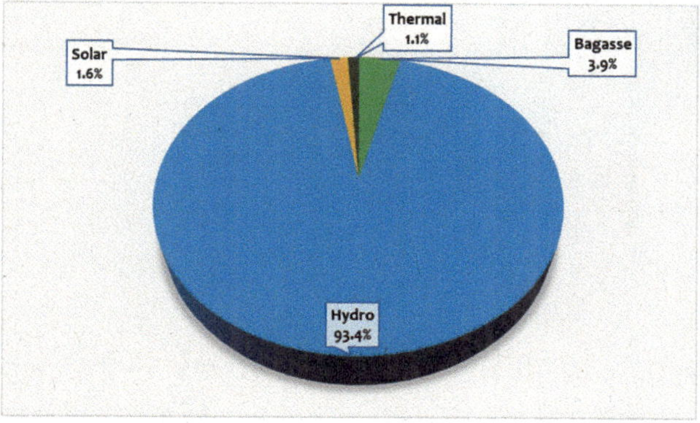

Fig. 9: Uganda's electricity generation from grid-connected power plants in 2023 (ERA, 2024).

In 2023, Kenya had a total installed power generation capacity of 3,321 MW, with 78 % derived from renewable sources such as geothermal, biomass, wind, solar, and hydropower. An additional 6 % came from the 500 kV *High Voltage Direct Current* (HVDC) link to Ethiopia, while the rest was produced from fossil fuel-based energy sources, as depicted in Fig. 10.

Kenya Power & Lighting Company is the utility responsible for most of Kenya's electricity transmission and distribution, while *Kenya Electricity Generating Company* (KenGen) is the country's biggest electric power generation company with a market share of 65 % (KenGen, n.d.) from its hydro, geothermal, thermal, and wind plants.

Fig. 11 shows the electricity generation mix for Kenya that totalled 13,289.63 GWh over the year ending June 2023. Geothermal output dominated at 6,035 GWh. A shortfall in hydro generation was offset by electricity imports via the Ethiopia–Kenya HVDC interconnection, supplemented by production from wind and solar energy (EPRA, 2023).

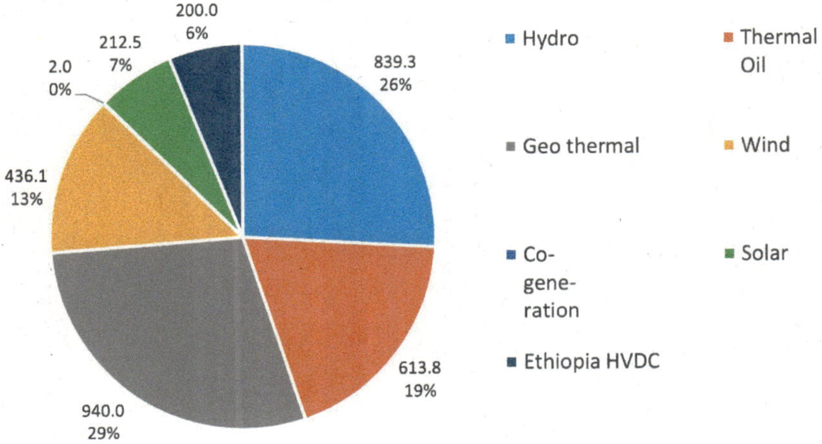

Fig. 10: Kenya's installed electricity capacity (in MW) and relative share of total capacity (in %) as at 2023 (Kenya National Bureau of Statistics, 2024).

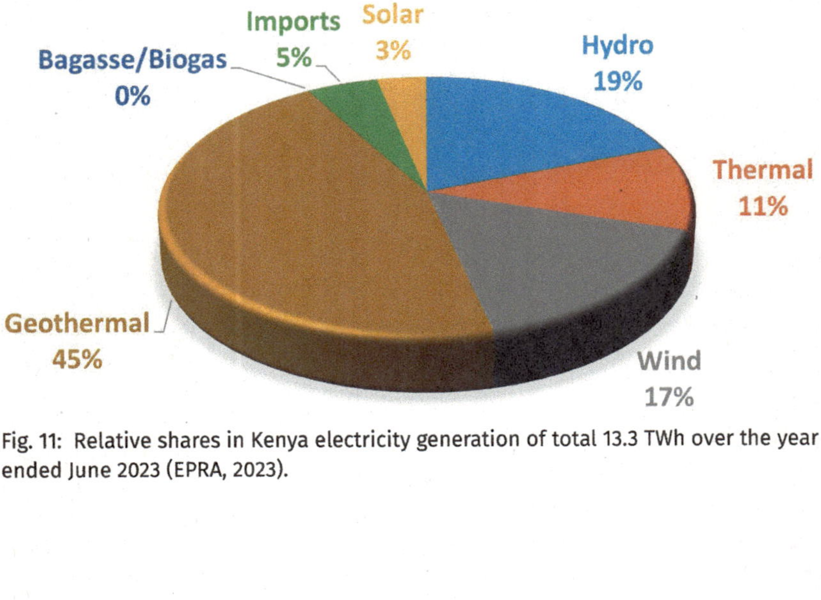

Fig. 11: Relative shares in Kenya electricity generation of total 13.3 TWh over the year ended June 2023 (EPRA, 2023).

3.1.5 Generation and Storage Options

Among East African countries, Kenya has made significant improvements in increasing power generation capacity from renewable energy sources. Currently, statistics indicate that renewable energy contributes over 80 % of the power injected into the Kenyan grid. Due to the intermittent power output of wind and solar, energy storage solutions are essential to facilitate the efficient adoption of renewable energy. As per *International Hydropower Association* (IHA) data for 2022, East Africa does not have pumped storage hydropower plants. The Kenyan government plans to incorporate *Battery Energy Storage Systems* (BESS) in its energy planning and policymaking. Such projects include a proposition of 250 MW in BESS facilities by 2024 and a gradual BESS capacity increase to 450 MW by 2036. These systems could deliver multiple benefits and solutions that can enhance the stability and reliability of the grid by addressing critical issues (Baru & Mutemi, 2023). Studies on the use of flywheels and supercapacitors as storage options in East Africa are generally limited to theoretical modelling and simulation (Mulumba & Farzaneh, 2023; Hendriks & Adonis, 2020).

3.1.6 Typical Energy Use Cases and Load Profiles in East Africa

Uganda's grid consumption in 2022 was 21.5 % by domestic users, 11 % by commercial users, 67.5 % by industrial users (medium, large, and extra-large), and 0.1 % for street lighting. Domestic consumption (60 %) in Kenya is primary for lighting purposes, which is slightly higher than the 50 % average for SSA household energy consumption attributed to lighting (Dominguez et al., 2021). In Tanzania, according to International Energy Agency (IEA) data, 44 % of total electricity consumption is contributed by the residential/domestic sector which constitutes the largest consumer. The second largest consumer with 32 % of the total is the commercial and public services sector and lastly the industrial sector at 21 % (Kofi et al., 2023).

While the public grid profile can (at least at some locations) also include considerable industrial consumers which have the highest consumption during daytime and less in the morning and evening hours, the typical microgrid profile has a relatively high proportion of domestic consumers who consume more during the morning and evening hours.

The average load profile for the Mpale microgrid in Tanzania is shown in Fig. 12 (Philipo et al., 2023). There is an evening peak between 7 and 9pm associated with increased domestic consumption due to household energy requirements such as lighting and domestic loads.

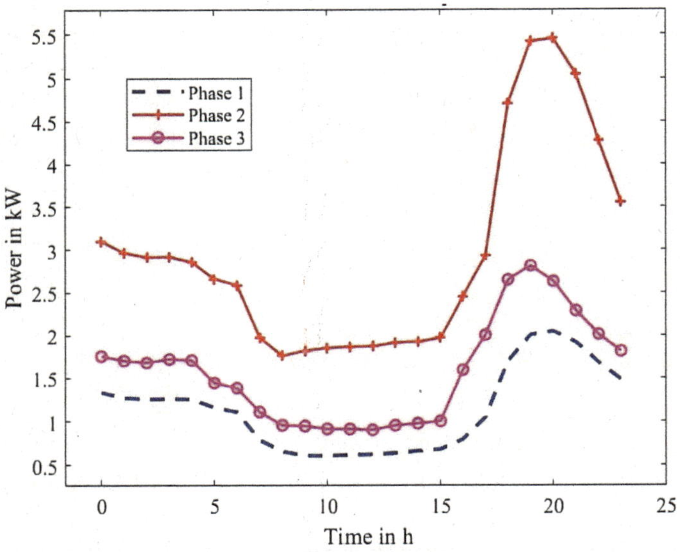

Fig. 12: Load profile of the three phases in Mpale microgrid in Tanzania (Philipo et al., 2023).

A *community microgrid* in Tanzania also displayed a similar load profile, as did *Senyondo microgrid* in Uganda, as shown in Fig. 13 and Fig. 14 (Kakande et al., 2024).

Fig. 12 and Fig. 13 show that load balancing on all three phases of microgrids and even the grid should be ensured, as uneven loading can cause increased outages on some phases (as was the case in the community microgrid), poor power quality, possible damage to equipment, and shortened equipment lifetimes.

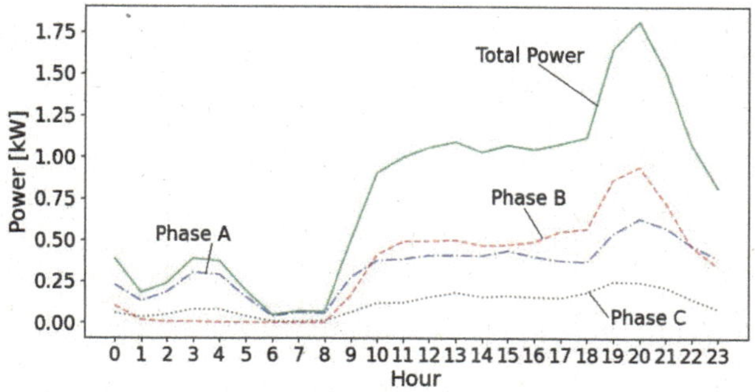

Fig. 13: Average load profile for community microgrid for 15–29 December 2022 (Kakande et al., 2024).

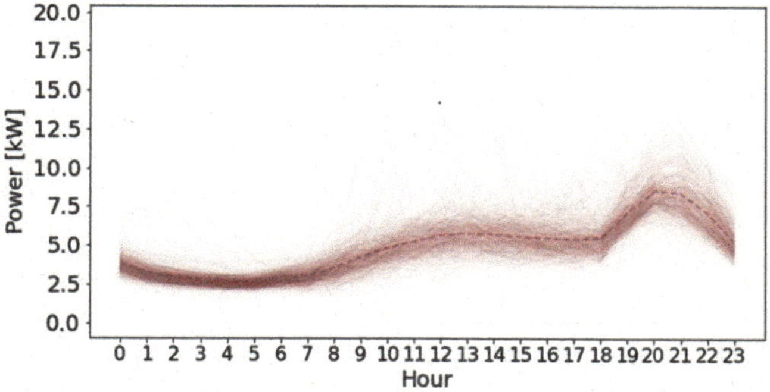

Fig. 14: Load profile for Senyondo microgrid from March 2022 to September 2023 (the dashed line represents the average values) (Kakande et al., 2024).

This general pattern is similar to that of the 11 Kenyan and Tanzanian microgrids presented by Williams et al. (2017) as well as the aggregated load profile of grid customers in Uganda (Seremba et al., 2024). This indicates that the region's load profiles are highly skewed by residential customers.

Some of the microgrids in Williams et al. (2017) as well as the grid consumption profile in Seremba et al. (2024) had an additional smaller daytime peak, which is attributable to commercial consumers.

All these load profiles indicate under-utilization of microgrid resources during peak PV production hours, which adversely impacts microgrid sustainability.

A study was thus done on the load profiles of East African microgrids that also explored the integration of electric two-wheelers and portable storage systems into a solar PV microgrid, the Senyondo microgrid on Bunjako island in Uganda. It showed that by incorporating e-mobility and portable storage, electricity access could be expanded and two-wheeler transport decarbonized. Modelling results showed a more than 98 % reduction in solar energy curtailment and an almost 60 % drop in the *Levelized Cost of Energy* (LCOE), with up to 73.27 tons of CO_2 emissions avoided annually. Additionally, the approach could support 160 off-grid customers yearly without expanding distribution infrastructure. This translates to improved financial and environmental sustainability of microgrids (Kakande et al., 2024).

Collaboration between microgrid developers and companies specializing in e-mobility and portable storage was recommended, alongside the establishment of conducive policies and regulations for e-mobility, as well as skilling of technical personnel to manage the microgrid and e-mobility and portable storage equipment and systems.

3.1.7 Recommendation

For microgrids in general and particularly solar PV systems where generation occurs during daytime, the lack of correlation between daytime generation and evening peak demand motivates the need for demand-side management measures that can reduce the evening peak loads and increase consumption during off-peak hours. This will lead to more sustainable microgrid operation due to the improved peak to average ratio and reduced requirements for storage to meet peak evening demand. The same would be useful for grid consumption with an evening peak.

Productive energy use is another recommended means of increasing demand during daytime hours, which can be done by availing income-generating appliances and machines such as refrigeration appliances, welding machines, carpentry tools, milling machines, etc. to microgrid customers at discounted prices and/or via an instalment payment plan. This facilitates the microgrid customers using electricity for more than simply phone charging and lighting and improves their ability to pay for electricity and to procure higher tier energy devices. However, it is important to ensure that the provided appliances are energy efficient.

3.1.8 Conclusion

Sustainable energy engineering in East Africa has good potential to support climate resilience through bridging the energy access gaps, hence driving economic growth. Leveraging the vast renewable energy resources together with innovative technologies to cater for their intermittency can ensure reliable and sustainable energy solutions. Utilizing proper technologies and ensuring necessary human capital for exploiting the available resources could make the region a global leader in sustainable energy; however, the role of all the different stakeholders should not be underrated.

Collaboration and partnerships, innovative business models, context-specific and context-appropriate energy use applications, technologies and loads that drive up demand during off-peak periods and reduce peak demand are all critical for ensuring sustainability of energy systems in a manner that encompasses financial, technological, social, and environmental aspects.

3.1.9 References

Al-Aghbary, M., Sobh, M., & Gerhards, C. (2022). A geothermal heat flow model of Africa based on random forest regression. *Frontiers in Earth Science, 10*. https://doi.org/10.3389/feart.2022.981899

Baru, E., & Mutemi, J. (2023). Kenya: The role of grid scale battery energy storage systems in achieving renewable energy goals. https://bowmanslaw.com/insights/the-role-of-grid-scale-battery-energy-storage-systems-in-helping-kenya-achieve-its-renewable-energy-goals/

Ddamulira, R. (2017). Climate change and energy in East Africa. *Development, 59*(3), 257–262. https://doi.org/10.1057/s41301-017-0101-1

Danny. (2004). East African community. *Wikimedia*, 1. https://commons.wikimedia.org/wiki/File:East_African_Community.JPG (28 August 2025).

Dominguez, C., Orehounig, K., & Carmeliet, J. (2021). Estimating hourly lighting load profiles of rural households in East Africa applying a data-driven characterization of occupant behavior and lighting devices ownership. *Development Engineering*, *6*, 100073. https://doi.org/10.1016/j.deveng.2021.100073

Electricity Regulatory Authority (ERA). (2023). *Electricity Supply Industry Performance Report for the Year 2022.* https://www.era.go.ug/download/electricity-supply-industry-performance-report-for-the-year-2023/?wpdmdl=7469&refresh=68d1225e5f7ae1758536286

Electricity Regulatory Authority (ERA). (2025). *Installed Generation Capacity.* https://www.era.go.ug/media/2025/07/Installed_Generation_Capacity_Q1_2025.xlsx

Electricity Regulatory Authority (ERA). (2024). *Energy Generated to the National Grid.* https://www.era.go.ug/index.php/stats/generation-statistics/energy-generated

Energy and Petroleum Regulatory Authority (EPRA). (2023) *Energy & Petroleum Statistics Report for the Financial Year Ended 30th June 2023.* https://www.epra.go.ke/energy-petroleum-statistics-report-financial-year-ended-30th-june-2023

Energy and Petroleum Regulatory Authority (EPRA). (2025). *Biannual Energy & Petroleum Statistics Report—Financial Year 2024/2025.* https://www.epra.go.ke/sites/default/files/2025-03/Bi-Annual%20Energy%20%26%20Petroleum%20Statistics%20Report%202024_2025.pdf

Global Wind Energy Council (GWEC). (2023). *The Status of Wind in Africa.* www.basi-go.com

Hendricks, T., & Adonis, M. (2020). Comparison of thermal, flywheel and supercapacitor energy storage for an African microgrid. In *Proceedings of the AIUE Congress 2020: International Conference on Use of Energy.* http://dx.doi.org/10.2139/ssrn.3735384

Hermann, S. Miketa, A. Fichaux, N. (2014). Estimating the Renewable Energy Potential in Africa. IRENA-KTH working paper. International Renewable Energy Agency, Abu Dhabi.

ICPAC. (2021). 10 climate change impacts in East Africa you didn't know about. *ICPAC.* https://www.icpac.net/news/10-climate-change-impacts-in-east-africa-you-didnt-know-about/

International Hydropower Association (IHA). Hydropower in Africa. https://www.hydropower.org/region-profiles/africa/

Kakande, J. N., Philipo, G. H., & Krauter, S. (2024). Optimized e-mobility and portable storage integration in an isolated rural solar microgrid in Uganda. *Solar*, *4*, 694–727. https://doi:10.1016/j.ecmx.2023.100381

Kariuki, B. W., & Sato, T. (2018). Interannual and spatial variability of solar radiation energy potential in Kenya using Meteosat satellite. *Renewable Energy*, *116*, 88–96. https://doi.org/10.1016/j.renene.2017.09.069

Katongole, D. N., Nyeinga, K., Okello, D., Mukiibi, D., Mubiru, J., & Kisira, Y. (2023). Spatial and temporal solar potential variation analysis in Uganda using measured data. *Tanzania Journal of Science*, *49*(1), 1–14. https://doi.org/10.4314/tjs.v49i1.1

KenGen. (n.d.). *Kenya Electricity Generating Company*. https://www.kengen.co.ke/index.php/who-we-are.html

Kenya National Bureau of Statistics. (2024). *2024 Economic Survey*. https://www.knbs.or.ke/reports/2024-economic-survey/

KFW Development Bank. (n.d.). *East African Community (EAC)*. https://www.kfw-entwicklungsbank.de/Global-commitment/Subsahara-Africa/East-African-Community/

Kichonge, B. (2018). The status and future prospects of hydropower for sustainable water and energy development in Tanzania. *Journal of Renewable Energy*, 1–12. https://doi.org/10.1155/2018/6570358

Kofi, A. P., Jammeh, K., & Odour, J. (2023). *An Outlook of Tanzania's Energy Demand, Supply and Cost by 2030*. Ed. 370 Working paper series, African Development Bank Group.

Magang, D. S., Ojara, M. A., Yunsheng, L., *et al.* (2024). Future climate projection across Tanzania under CMIP6 with high-resolution regional climate model. *Scientific Reports, 14*, 12741. https://doi.org/10.1038/s41598-024-63495-w

Mulumba, A. N., & Farzaneh, H. (2023). Techno-economic analysis and dynamic power simulation of a hybrid solar-wind-battery-flywheel system for off-grid power supply in remote areas in Kenya. *Energy Conversion and Management: X, 18*, 100381, https://doi:10.1016/j.ecmx.2023.100381

Othieno, H., & Awange, J. (2016). Energy resources in East Africa. In: H. Othieno & J. Awange (eds.), *Energy Resources in Africa: Distribution, Opportunities and Challenges*. Cham: Springer. https://doi.org/10.1007/978-3-319-25187-5

Parry, M., & Palutikof, J., et al. (2007). *Climate Change 2007: Impacts, Adaptation and Vulnerability.* Cambridge: Cambridge University Press.

Philipo, G. H., Kakande, J. N., & Krauter, S. (2023). Demand-side-management for optimal dispatch of an isolated solar microgrid. *IEEE AFRICON Conference*, 1–6. https://doi.org/10.1109/AFRICON55910.2023.10293343

Seremba, N. E., Ssemakula, F., Namaganda-Kiyimba, J., & Kakande, J. N. (2024). Integration of the centralized grid and decentralized renewable energy off-grid systems: A techno-economic analysis. *2024 IEEE Conference on Technologies for Sustainability (SusTech)*, 257–263. https://doi.org/10.1109/sustech60925.2024.10553436

Tanzania Power Production and Demand—2024 Update. (n.d.). https://www.tanzaniainvest.com/power

Williams, N. J., Jaramillo, P., Cornell, B., Lyons-Galante, I., & Wynn, E. (2017). Load characteristics of East African microgrids. *Proceedings—2017 IEEE PES-IAS PowerAfrica Conference: Harnessing Energy, Information and Communications Technology (ICT) for Affordable Electrification of Africa, PowerAfrica 2017*, 236–241. https://doi.org/10.1109/PowerAfrica.2017.7991230

World Meteorological Organization (WMO). (2025). WMO confirms 2024 as warmest year on record at about 1.55°C above pre-industrial level. https://wmo.int/news/

media-centre/wmo-confirms-2024-warmest-year-record-about-155degc-above-pre-industrial-level

Xie, B., & Xie, B. (2024). Assessing the impact of climate policy on energy security in developed economies. *International Review of Economics and Finance*, *90*(1), 265–282. https://doi.org/10.1016/j.iref.2023.11.006

3.2 Designing and Constructing Microgrids for Local Communities

Stefan Trittler, Joachim Böcker, Henrik Bode

There are many factors that influence the performance of an off-grid photovoltaic (PV) system used in microgrids, many of which need to be considered before the system is commissioned, others which are relevant during operation. Therefore, we need to assess what technical or design parameters need to be considered to achieve a successful operation of a microgrid for local communities.

The system must meet the intended use and expectations based on the initial requirements and design assumptions. Due to its highly project-specific nature, its operational success depends on the thoroughness of the needs assessment and design work, which enables the operation teams to operate the PV off-grid system effectively. The design of a system is influenced by operational requirements, while operational success is equally dependent on an effective design. These two factors are interdependent and require thorough assessment during the design process. In certain scenarios, an over-performing design may be required to meet certain expectations, while an under-performing design could limit some operational capabilities. However, cost considerations often impose constraints on ideal operation, which can lead to system failures or limitations.

Our goal in this chapter is to explain the factors that contribute to the successful operation of an off-grid PV system. We have divided the considerations into three phases: the early need assessment and design phase; the implementation phase; and the operational phase with a focus on safety.

Designing a robust off-grid PV system requires an approach that goes beyond technical specifications. There is a strong emphasis on reliability, resilience, and component selection based on proven performance, as well as the inclusion of redundancy whenever possible. An important aspect of the design is the efficiency, scalability, and ease of maintenance, as well as the seamless integration of the components and consideration of the environment. Lastly, long-term success requires cost effectiveness, regulatory compliance, and adaptability to local conditions. Apart from all these engineering requirements, many social and economic factors play a role, like the engagement of users and beneficiaries, business models and tariffs, communication and expectations management, and so on, which will be discussed in subsequent chapters.

3.2.1 Important Design Considerations for Microgrids

The design of an effective and efficient PV off-grid solution is a complex process that requires careful consideration of a number of factors. This section will examine the critical aspects of load evaluation, sizing and lifetime expectations, technological selection, and electrical design.

3.2.1.1 Load Evaluation and Needs Assessment as Main Prerequisite

In the design phase, the evaluation of loads is not the sole focus; rather, the objective is to define the application with precision. The application scope of an off-grid PV system extends beyond mere technical considerations to encompass practical use in a variety of contexts, including the provision of electricity to remote communities, communication networks, research facilities, and other remote loads. The adaptation of the system design to meet the requirements of these applications requires an understanding of the diverse energy needs, usage patterns, and critical performance parameters that inform the decision-making process. An analysis of historical data on energy consumption and load requirements is instrumental in accurately estimating future loads. This information is crucial for the sizing of battery systems and the determination of system capacity to ensure reliable and optimal energy delivery without overload or underperformance.

3.2.1.2 Demand-Side Management (DSM)

In addition to load and demand assessment, it is important to identify critical loads and whether certain loads can be deferred or included as an option. On the one hand, load shedding can be implemented as a passive safety feature to protect critical loads when the power system is under-performing, preventing a complete system shutdown. On the other hand, it can be used as an active approach to managing loads based on power availability and demand, known as Demand-Side Management (DSM). DSM strategies can be controlled automatically, manually or through a semi-automated system that guides user decisions.

While Chapter 3.3 explores various DSM options in more detail, it is important to emphasize here that these considerations need to be addressed at the design phase. This will ensure that the system is sized appropriately and that the DSM system can operate effectively in the future.

3.2.1.3 Location and Environmental Impact as Important Preconditions

A thorough examination of the environmental factors that are critical to optimal system performance is essential in the context of off-grid PV systems, with particular attention paid to the pre-conditions and location. The intensity of solar irradiance is of particular significance, with the objective of locating areas with high irradiance in order to maximize the energy output of the PV panels. Conversely, extreme temperature fluctuations have a significant impact on the performance and longevity of the batteries. Furthermore, altitude and proximity to rivers, seawater or water runways during heavy rains introduce additional complexities that influence atmospheric conditions and necessitate the implementation of specialized protection measures.

In the context of assessing the potential impact of installations in areas such as nature reserves or high-risk regions, it is necessary to consider how the operation or construction of the system may affect these sensitive locations. For instance, the implementation of off-grid PV systems in conservation areas may necessitate adherence to particular environmental regulations to prevent disruption to local ecosystems. Similarly, in areas prone to natural disasters, additional protective measures may be required to mitigate potential damage to the system and the surrounding environment.

Furthermore, the availability of qualified personnel on site is fundamental for the timely maintenance and troubleshooting of the system, thus reducing technical issues and ensuring optimal system performance. In the event that the remote side is not easily accessible, it is essential to consider the incorporation of design elements, such as redundancy, to ensure the reliability of the system. Additionally, a reliable Internet connection is required to facilitate remote monitoring and prompt response to system issues or emergencies.

3.2.1.4 Sizing of the System Considering Lifetime Expectations

The focus of subsequent analysis will be on sizing considerations and the establishment of lifetime expectations. This necessitates the precise estimation of the power and energy rating of components, including PV panels, electronic devices, and battery banks, to fulfil identified load requirements. It is of key importance to make appropriate sizing decisions in order to optimize energy generation, storage, and distribution within an off-grid PV system.

As the battery is a crucial component in the off-grid PV system, the determination of an appropriate battery size is a core aspect of the design. This process entails calculating the required storage capacity to effectively contain the surplus energy generated by the PV panels, particularly during periods of peak solar irradiance. It is of absolute necessity that the batteries be accurately sized in order to guarantee that they have the sufficient capacity to meet the energy requirements of the application, particularly in scenarios with limited or no sunlight. A primary objective is to reduce reliance on external sources, such as diesel generators, by emphasizing autonomy, particularly during periods of low renewable energy availability or system maintenance.

The following factors must be considered:
Depth of Discharge (DoD): It is often assumed that certain battery technologies can be fully discharged, but this depends on the specific chemistry or limits set by the manufacturer to improve longevity and performance. In general, discharging a battery to an extremely low level of capacity or charging it to a very high level can adversely affect its overall life.

Number of cycles: The number of cycles is dependent on a number of factors, including DoD, C-rate, temperature, and other variables. Performance varies between chemistries and is also contingent on the quality assurance measures employed by the supplier.

C-rate: Lithium-ion batteries have a higher C-rate than many alternative battery technologies. However, there are notable discrepancies in their chemical composition, with suppliers often distinguishing between high power and high energy cells. While applications such as automotive and consumer electronics often require high C-rates (exceeding 2C), most energy demands in off-grid stationary PV systems occur at rates below 1C. High power batteries offer potential advantages in certain applications, but their suitability often requires careful evaluation. This is because the high C-rate they offer may not be maintained at full capacity or may be limited by internal temperature limits. In contrast, high energy batteries tend to be less complex in design and more cost effective, making them a practical choice for many scenarios. In addition to lithium-ion batteries, lead-acid batteries continue to play an important role in off-grid PV systems, where they have dominated the sector for years. These batteries typically operate at a C-rate of less than 0.5C, with system designs typically based on a C-rate of C_{10}.

Furthermore, consideration is given to establishing realistic expectations regarding the lifetime of the battery, particularly in the context of battery warranties and the conditions associated with them. This includes a detailed examination of warranty terms, considering factors such as cycle life, DoD capabilities, and the manufacturer's stipulations. The interplay between sizing decisions, oversizing considerations, and lifetime expectations ensures that the design aligns with both performance goals and warranty conditions, promoting a robust and reliable off-grid PV system.

A robust and efficient off-grid PV system design philosophy is based on the above principles for battery design, as well as the appropriate sizing of the PV system and the related electronics.

Depending on the location and irradiation forecasts, the PV system orientation, inclination, size, and structural setup needs to be defined. Usually simulation tools help to define the system with optimal PV yield, but load patterns can be considered here as well. As PV panels have become comparably cheap, an east-west orientation or an oversizing of PV might make sense. In the end, the PV system has to provide the needed energy demand including system losses—in most cases with 100 % renewable energy fraction, the PV system is oversized in high irradiation times as it is designed to meet the load in low irradiation times. Batteries are used to store energy for a few days but not for seasonal variations.

PV chargers or inverters need to be sized according to the PV system while battery inverters are dependent on the peak loads. Usually in 100 % renewable energy systems, the PV size is much higher than the peak load, but as these loads might occur during the night, the battery has to power them. Also, for system stability reasons, a certain share of battery inverter capacity is required as they are the grid-forming device in a 100 % renewable energy system.

As microgrids for local communities are usually far from any infrastructure and not connected to public grids or regular fuel supplies, they are commonly designed for a PV only operation (100 % renewable energy) while allowing for DSM in case of low performance. Alternatively, a backup diesel generator could be considered to reduce oversizing of PV and battery to supply special loads or overcome high demands or low PV yields, but this approach is very project-specific.

3.2.1.5 Selection of Technology and Price vs Quality Consideration of Suppliers

The success of an off-grid PV system is dependent upon a comprehensive evaluation of the available suppliers and technologies. This assessment prioritizes technological reliability, with a particular focus on the performance history, durability, and adaptability to various conditions of the technology in question. Furthermore, it is vital to ascertain the dependability and accessibility of technical assistance from the selected suppliers, as this will facilitate the expeditious resolution of any issues that may arise and ensure the uninterrupted operation of the system. Portability considerations are of central importance, particularly in the case of systems that require mobility or reconfiguration.

It is also essential to verify the functional compatibility of components from different manufacturers, particularly in conjunction with lithium-ion batteries. Obtaining approval from all relevant parties will ensure that the warranty remains valid. Moreover, a careful selection and adaptation of the appropriate charging/discharging algorithm is essential. Charge controllers are responsible for regulating the charging process, preventing potential issues such as overcharging or over-discharging, and ensuring the system's stable performance. Similarly, the selection of a pure sine wave inverter is important to ensure compliance with sensitive electronic devices. Moreover, it is of great value to adapt the system to specific applications, which mitigates the risks associated with potential mismatches between equipment and application requirements. Through a seamless integration of all components with the Energy Management System (EMS) and the Battery Management System (BMS) facilitating robust communication and control mechanisms, potential system failures (e.g., inaccurate readings or improper voltage regulation) are reduced and overall performance is improved.

The characteristics of different battery technologies, especially different lithium-ion chemistries, vary considerably and have a direct impact on factors such as energy density, lifespan, safety, and performance under varying load or environmental conditions. It is therefore vital to select an appropriate battery that is aligned with the specific requirements of the off-grid PV system and to select suitable power electronics and BMS in order to optimize its functionality, safety, and longevity.

Once a product has been selected, simulations can be employed to facilitate a re-evaluation of the system design. This process may necessitate modifications to accommodate precise product specifications, thereby ensuring seamless integration and compatibility among the selected components. The utilization of simulations allows for a proactive approach, whereby potential constraints that may have emerged during the product selection process can be identified. By incorporating simulation tools, the design can be refined to address these issues while aligning with the overarching objectives of the off-grid PV system.

3.2.1.6 General System Design Looking at Side Aspects

A comprehensive examination of the general configuration of the off-grid PV system reveals a number of factors that warrant particular attention.

In examining the complexities of electrical designs, particular emphasis is placed on the DC bus voltage also in light of the selected equipment and load requirements. Factors such as voltage levels, power conversion efficiency, and system stability are subjected to close examination to ensure that the electrical design fully exploits the potential of the selected technology within the context of the off-grid PV system.

Another consideration is the safety of the O&M personnel. In the case of extra-low voltage systems (less than 60 V DC), there is no risk of electric shock, whereas high voltage systems (typically 120–1500 V DC) necessitate the involvement of specially trained personnel and the use of appropriate equipment.

It is imperative to examine strategies for effective cooling mechanisms in order to maintain optimal operating temperatures for system components, thereby mitigating the risk of overheating and performance degradation. Environmental factors, including meteorological conditions and climate variability, are evaluated and protective measures are implemented to ensure system security.

Remote monitoring systems that provide real-time insight into the operational status and overall health of the off-grid PV system can provide important information for successful operation. It is crucial to monitor the state of charge (SOC) of the batteries in off-grid PV systems, as different battery technologies have varying optimal SOC ranges. For lithium-ion batteries, maintaining consistently high SOC levels can significantly reduce their lifespan. In contrast, lead-acid batteries benefit from being at full or near full charge to maintain health and longevity but are severely affected by low charge—they are best kept above 50 % SOC. Accurate SOC assessments enable users to optimize energy storage levels, validate initial forecasts, and make operational adjustments to improve system availability, longevity, and robustness.

The strategic placement of the off-grid PV system is of critical importance, with the mitigation of fire risk and compliance with safety protocols representing key priorities. Through careful planning, the potential for fire risks can be reduced and compliance with safety regulations ensured. In regions prone to flooding, the waterways must be subjected to careful assessment to identify any vulnerable locations that could damage the off-grid PV system during floods.

The incorporation of redundancy strategies represents an important element for the improvement of safety in an off-grid PV system. This includes the use of redundant components or systems that can seamlessly take over in the event of a failure, thereby minimizing the likelihood of critical failures and ensuring continuous operation even under challenging conditions.

3.2.2 Development and Applications of Power Electronics

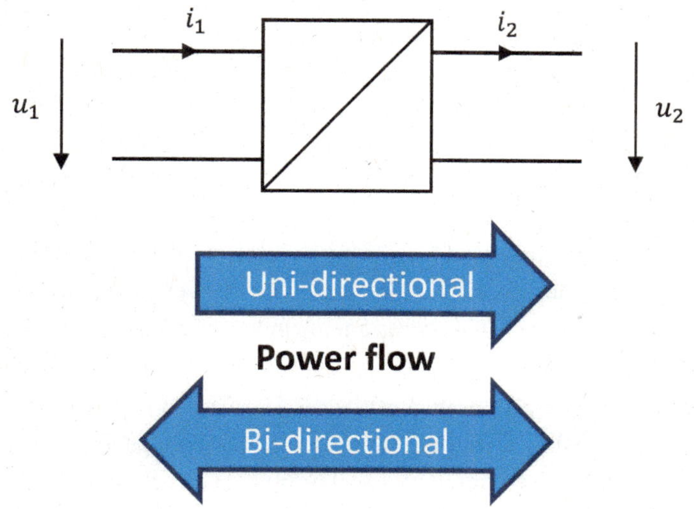

Fig. 15: General circuit symbol of a converter.

Power electronics has become a key technology in the field of the generation, distribution, and consumption of electrical energy. It allows electrical energy to be converted into a form that best suits the intended purpose. These are conversions from AC to DC (rectification), DC to AC (inversion), as well as conversions between different voltage levels. However, the conversion between various AC voltage levels can also be performed by means of power electronic converters, which offer some advantages over the classical transformer. The power flow through these converters is in most cases unidirectional, i.e., from the input to the output as is the case with, e.g., PV inverters, or bidirectional, which is necessary to connect a battery as backup for off-grid operation (see Fig. 15 and Fig. 16).

Conversion	to	
from	**DC**	**AC**
DC	DC-DC converter	Inverter
AC	Rectifier	AC-AC converter

Fig. 16: Wording for power electronic converters.

However, with the help of power electronic converters, the form of electrical energy can be adapted to the needs of the involved components, particularly those which appear as components in a microgrid.

Examples:

- The electrical power grids are worldwide AC grids. On the larger power scale, it consists of three phases on various voltage levels between 10 and 380 kV at frequencies of 50 or 60 Hz (so-called medium and high voltage). The consumers, however, are usually supplied with low single-phase voltage in the range of 110 to 240 V or three-phase voltage of 190 to 415 V, respectively. Note that the given numbers are nominal values, depending on the countries' standards. In Kenya, Tanzania, and Uganda, the nominal voltage is 230/240 V at a frequency of 50 Hz. The actual voltage values, however, can fluctuate considerably, e.g., up to +/- 10 % in Germany, but even more so in regions of weak supply.
- The worldwide market of household appliances (fridges, lighting, TV, cookers, etc.) provides standard products for 230/240 V AC, 50/60Hz

(or for 110 V). Due to mass production, the appliances come at favourable cost, though the standardized voltage may not be optimal for the particular application, for example:

- A single lighting LED has a rated DC voltage of only about 3 V, several in series of about 10 to 15 V, but they are also available for 230 V AC sockets.
- A PV panel generates a DC voltage of about 30–45 V. Depending on the number of panels connected in series, the voltage of a PV array results in some 100 volts but would usually not exceed the threshold of 1000 V.
- Battery stacks for supporting off-grid operation come with DC voltages of 50 to some 100 V, depending on the number of cells in series and the state of charge. So, it is a similar voltage range as with PV arrays.

Thanks to considerable advances in power electronics devices in recent decades, the conversion between the different voltage levels and voltage types (DC vs. AC) is today possible with electronic converters at reasonable cost and with only few losses. For example, PV inverters that convert the DC voltage of a PV array to the AC grid voltage produce only 2–3 % losses, i.e., they have an efficiency of 97–98 %.

Another important point is the control of the power flow. Unlike the classical passive diode bridge rectifier or the common transformer, where the power flow is uncontrolled and the result of the connected supplies and loads of the AC and DC sides, modern power electronic converters can actively control the output voltage, current, or power in an intelligent way. In this manner, even sophisticated tasks such as maximum power point tracking, short circuit protection, and grid code implementation are possible.

The key elements of today's power electronic converters are electronic switches which are semiconductor transistors and diodes. Today, silicon semiconductors are dominant, but novel silicon carbide and gallium nitride devices are entering the market, allowing even lower losses, and consequently further shrinking of the converters' mass and volume.

Simply put, by turning on and off these semiconductor switches (a basic switch is shown in Fig. 17) at a rather high switching frequency of usually

10–500 kHz, the input voltage is chopped and the desired output voltage is composed out of these chopped input voltage fragments, which is called *Pulse Width Modulation* (PWM). As an example, Fig. 18 shows the circuit of a PV inverter, and the associated PWM scheme is given in Fig. 19.

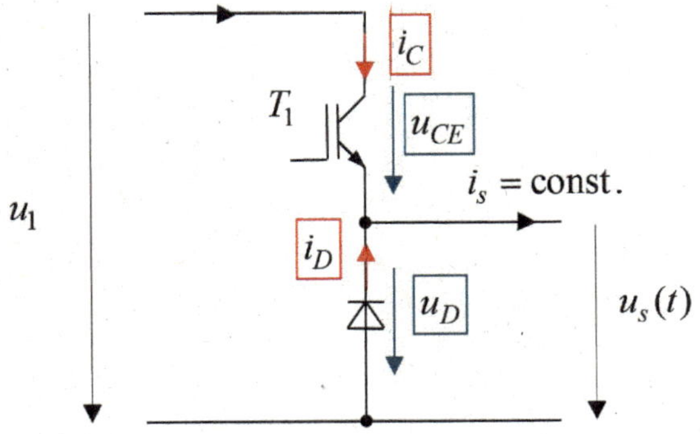

Fig. 17: Basic switch element of a DC-DC converter consisting of a transistor and a diode.

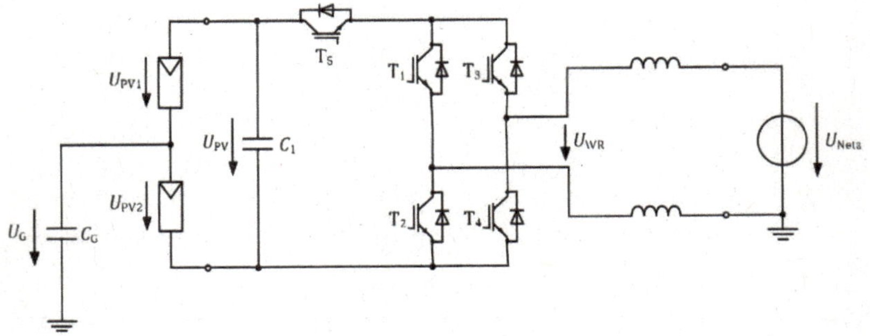

Fig. 18: Circuit diagram of a single-phase PV inverter (shown with a grid connection).

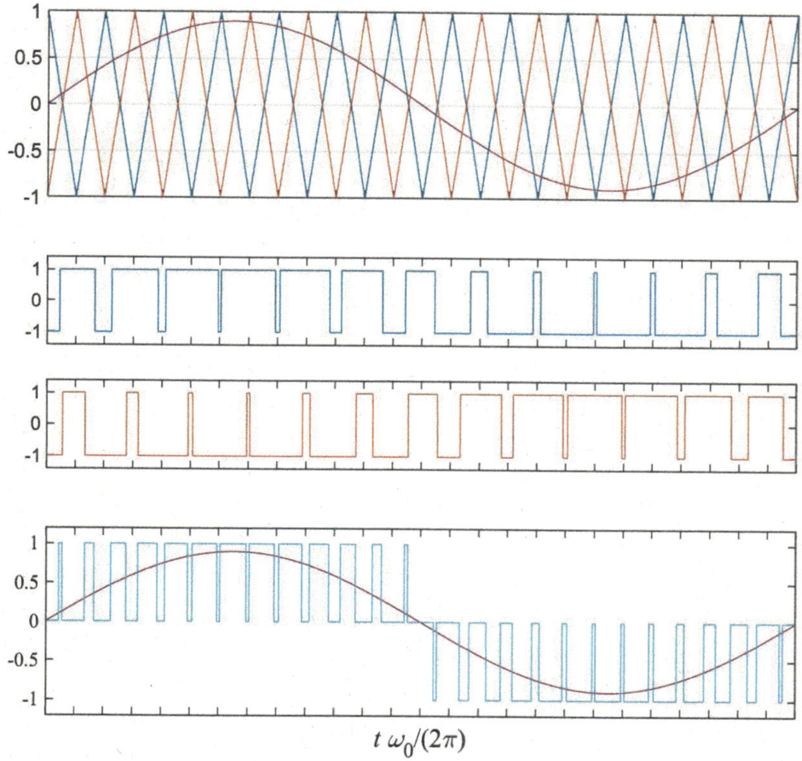

Fig. 19: Principle of Pulse Width Modulation (PWM) for a PV inverter. Plot 1 (upper): Desired sinusoidal grid voltage and the so-called triangular-shaped modulation carriers. Plots 2 & 3: The switching commands for the two inverter legs, resulting from comparison of grid voltage and the respective modulation carrier. Plot 4 (lowest): The »chopped« relative inverter output voltage in comparison with ideal sinusoidal curve.

After these remarks, it may look like an ideal setting in which to establish a microgrid. In principle, this is true, but the situation is less straightforward than it initially seems.

The situation is that general-purpose converters are rarely available on the market. Most converters are dedicated to very specific applications. For example, to connect a PV array to the AC grid, an inverter is necessary, inverting the DC voltage to single-phase or three-phase AC voltage. However, it is a specialized PV inverter. Although a similar inverter is also needed to feed an electric AC motor from a DC battery supply, one cannot use an PV inverter for this purpose; for that, one would have to use specialized motor inverters.

The reason is that the power electronic design has to be adapted to the very special specification of the application in terms of rated voltage, rated power, and operation modes. Furthermore, the power electronics devices comprise not only the power electronic circuitry but also the control which is even more specialized to the needs of the particular application. In the case of the PV inverter, it includes maximum power point tracking on the DC side and grid code aspects, e.g., control of reactive power, on the AC side. It is obvious that such a design is complex and pays back only after a sufficiently large number of produced pieces.

When it comes to the design of a novel individual microgrid, this means that one has to use components that are available on the market. That was also an issue in our project. One of our goals was to investigate how neighboured microgrids could be connected for mutual benefits also during grid off-times. From a technical viewpoint, DC coupling would have some benefits, but because the required components were not available on the market, we focussed on AC coupling.

Due to limited resources for the project, it was also not possible to establish really separated microgrids. In order to simulate such a situation, the consumers on the LWAK campus were divided into three subgrids—according to the three phases—which can operate separately in off-grid mode but can also be coupled with each other or with the supply grid. The implemented configuration is shown in Fig. 20.

The system is equipped with a large number of contactors to investigate various configurations as:

- Normal grid-connected operation mode without PV and diesel generator
- Supply only with diesel generator without PV and without supply grid
- Battery supply in combination with or without PV
- Various mixed modes

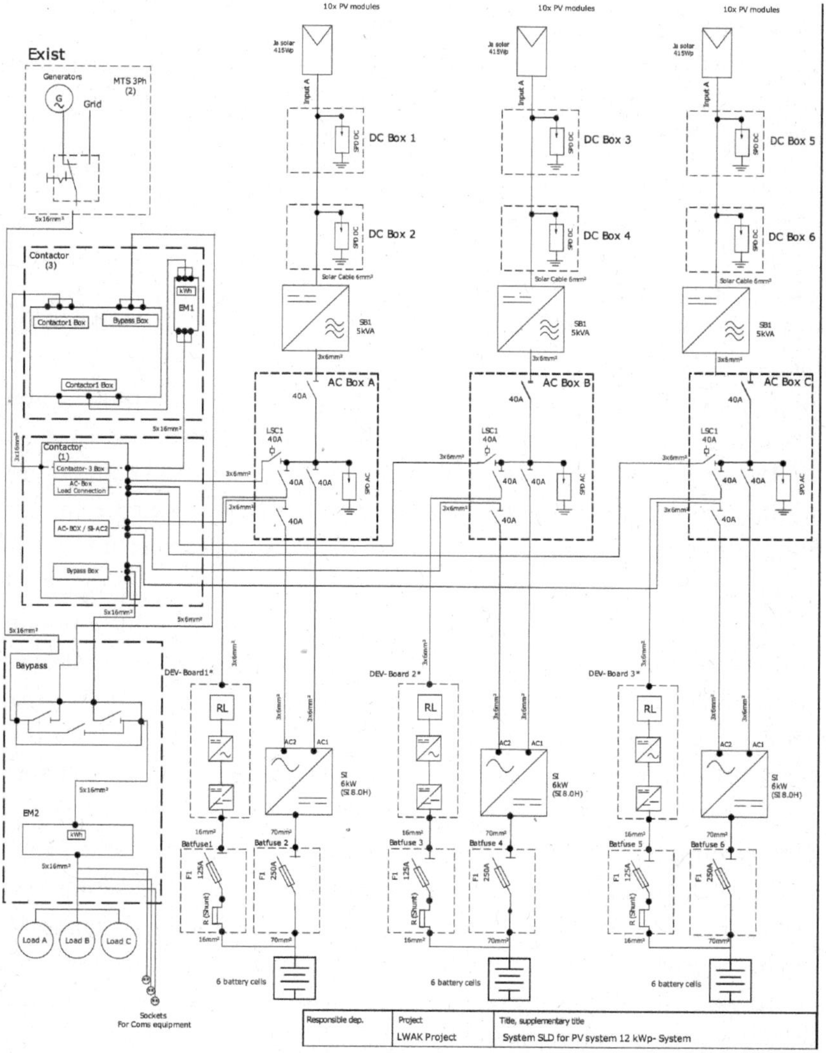

Fig. 20: Circuit diagram of the LWAK microgrid system, divided into three subsystems: A, B, and C.

It should be noted that the size of the PV array is not large enough to run the LWAK campus in a fully self-sufficient mode. This was originally intended but was also not achievable due to limited project budget.

3.2.3 Implementation of a Microgrid

This sub-chapter explores the crucial phase of implementing PV off-grid systems to power microgrids in local communities. It covers the essential steps of transportation, installation, testing, commissioning, and training, emphasizing their collective impact on the system's functionality and longevity.

3.2.3.1 Transportation of Critical Equipment to Remote Sites

In the transportation phase, strategic planning is imperative to facilitate the safe and efficient movement of the equipment to the respective location. Some roads to remote locations cannot be accessed during the rainy season or require 4x4 vehicles or other means for transportation.

The safe transportation of equipment like PV panels and batteries necessitates a comprehensive logistics assessment that includes the engagement of reputable transport providers with the requisite expertise to handle the sensitive components.

Compliance with the safety regulations that govern the transportation of hazardous materials, as well as the industry-wide standards that have been established for this purpose, are an indispensable aspect of the logistics process.

The packaging requirements assume great significance, necessitating robust and protective packaging to shield the equipment from potential damage during transit. Particular emphasis is placed on the implementation of safety measures during the transportation phase. This encompasses the secure fastening of components to prevent movement, the accurate labelling of packages, and the establishment of clear emergency response protocols in the event of unforeseen incidents. This comprehensive approach to transportation guarantees the equipment's save arrival at the designated off-grid PV system location.

Furthermore, extended transportation, particularly to distant locations such as Africa, may require recharging of the batteries due to prolonged periods without immediate installation. The duration and impact of this situation can vary depending on the terms of the warranty. The time spent in transit, including periods spent awaiting clearance through customs, further exacerbates the issue. In areas that are particularly remote, the logistics of transporting, installing, and maintaining batteries become even more challenging. It is essential to consider the effects of calendric ageing in such scenarios, as this can affect the batteries' performance over time. The self-discharge of the batteries also needs to be considered, as they may need to be recharged within a certain time during the transport and implementation phase.

3.2.3.2 Training as an Important Factor for Successful Operation

In the initial training phase, the emphasis is placed on the installation of the off-grid PV system by proficient installers. Subsequently, the focus shifts to operators and maintenance personnel who need to be equipped with the requisite knowledge and skills for the efficient operation and maintenance of the system.

This comprehensive training programme encompasses guidance on system operation and routine maintenance procedures, including essential techniques for starting and stopping the system, interpreting system indicators, and developing troubleshooting skills to promptly resolve common issues. Furthermore, the training encompasses ancillary equipment such as air conditioning and fire detection and suppression systems, thereby ensuring that personnel are adequately prepared to manage all aspects of the system's functionality and to respond to emergencies and potential hazards associated with electrical power in general or lithium-ion batteries, for example. By ensuring that personnel possess the requisite knowledge and skills, the training phase makes a substantial contribution to the ongoing reliability and successful operation of the off-grid PV system.

3.2.3.3 Correct Installation as a Crucial Factor

Installing an off-grid PV system requires a variety of skills and many different trades coming together. Depending on where the PV system is installed (roof, free-field, etc.), structural assessments or foundation designs need to be undertaken. Also, depending on the available housing options for the battery and electronics, extra construction work or the use of containerized systems might be required.

All the various components have different requirements concerning installation, and mistakes can have a major impact on operation, which is why well-trained and skilled installers are vitally important in creating a successful operating microgrid. Again, the battery, as one of the most crucial components in the system, requires the most care. Different battery types have different requirements and risks that need to be addressed during installation. The first step is to ensure that the technical room is adequately ventilated, or even cooled if necessary. It is of chief importance to exercise the utmost care when handling batteries and to strictly follow safety guidelines. The mishandling of batteries during installation often results in mistakes such as incorrect wiring connections, inadequate insulation, or insufficient securing of the batteries in their housings or racks. These mistakes can lead to operational inefficiencies, safety hazards, or even premature battery degradation. Given the sensitivity of batteries to environmental conditions and their intricate design, installation errors can have a significant impact on the overall performance and longevity of the off-grid PV system.

3.2.3.4 Testing and Commissioning as Final Step before Operations

Prior to commencing the testing phase, it is imperative to undertake a comprehensive examination of the installation. When starting up the system, all installation work should be completed and verified. After the first operational checks and setting of parameters have been undertaken, the system can be tested. This means checking the full operation of the PV system and the battery; it is advisable to carry out voltage checks to ensure that all systems work as intended. In addition, a capacity test of the battery could be performed at this stage. This process assesses the actual energy storage capacity of batteries in comparison to the specifications outlined by the manufacturer. Furthermore, diagnostic techniques such as thermal imaging can be employed. Through these rigorous testing protocols, any anomalies or deviations are identified and addressed, thus ensuring the reliability and efficiency of the system.

Upon successful completion of the testing phase, the commissioning phase begins, which signifies the formal commencement of the off-grid PV system. This phase entails the configuration of specific settings on the BMS or other control systems, ensuring that parameters align precisely with the system requirements. Concurrently, a particular validation process is conducted to guarantee seamless communication interfaces between the batteries, other components of the PV system, and the loads. This step is vital for the cohesive integration of the off-grid PV system into the broader microgrid, paving the way for their reliable contribution to the overall functionality of the system.

3.2.4 Safety Aspects

Navigating the complexities of environmental considerations is crucial for ensuring the longevity and optimal performance of the PV off-grid systems.

Effective temperature control of facilities in remote areas with high ambient temperatures is of particular importance. However, the intricacy involved and dependence on expert technicians for active cooling systems can present a significant challenge, potentially resulting in malfunctions or thermal complications such as uncontrolled fires. Depending on the battery technology used, the local conditions, and the system setup selected, different precautions, e.g., on cooling or fire protection need to be considered. For lithium-ion battery systems, this includes incorporating fire extinguisher systems, using fire resistant enclosures, installing smoke detectors, and establishing emergency shutdown procedures. These measures are essential to manage the specific risks associated with lithium-ion batteries, which are more prone to thermal runaway.

Although the risk of fire is generally lower for lead-acid battery systems, it is still important to take appropriate safety measures. These include ensuring proper ventilation to prevent the build-up of hydrogen gas, which is flammable, and using protective enclosures to protect against short circuits.

To ensure the safety of stationary batteries in off-grid PV systems, it is essential to implement a set of comprehensive measures designed to mitigate potential risks: lithium-ion batteries need the implementation of an advanced BMS to control parameters and enforce strict safety measures at the battery cell and module level to prevent malfunctions and optimize performance. Although lead-acid batteries do not typically require an advanced BMS, regular monitoring and maintenance is critical. This includes ensuring proper ventilation to prevent the build-up of hydrogen gas during charging, which can pose a risk of fire or explosion. In addition, maintaining an appropriate charging regime, avoiding deep discharges, and checking electrolyte levels (for flooded lead-acid types) are necessary to ensure safety and prolong battery life.

From a holistic perspective, there is a particular emphasis on compliance with local regulations and building codes. This is to ensure comprehensive safety and regulatory compliance. Despite this, it is important to note that within this broader context, the implicit consideration is to align all equipment with pertinent standards and regulations. This approach can enhance the safety and reliability of the system.

3.2.4.1 Health and Safety Measures

The well-being of personnel involved in the installation, operation, and maintenance of off-grid PV systems is of the utmost importance. Health and safety measures, including comprehensive training protocols, the use of protective equipment, and adherence to safe handling guidelines, are an integral part of the overall risk mitigation strategy. Different types of system design and the use of different equipment present different risks, and the design or installation methods may impose specific health and safety requirements.

For example, lead-acid batteries require proper ventilation, and the handling of their hazardous acid requires the use of protective equipment. Older systems predominantly use lead-acid batteries operating at extra-low voltages (<60 volts). Modern lithium-ion systems, on the other hand, typically operate at higher voltages, resulting in greater safety complexity. This increased complexity often requires the involvement of certified electricians to ensure safe installation and maintenance.

3.2.5 Realization of Microgrid Demonstrator in Lwak, Kenya

To flesh out the developed theories with empirical work, the plan of creating an on-site laboratory was born. This field laboratory has been created to support the research in three different ways. The first and most obvious way is to test the previously developed technical concept. Second, this work enables an analysis of how creating a community-driven microgrid affects the socio-economic behaviour of the selected group. Third, a public-owned microgrid can serve as both a teaching demonstrator and a good practice multiplicator, not only for the convent, the attached schools, and the local community of the nearby village, but also for teaching local technicians with the training concept of SMA sunbelt.

3.2.5.1 Convent of the Franciscan Sisters of St. Anna in Lwak, Kenya

St. Anna's Convent as a location for this setup was selected to maximize its impact, feasibility, and long-term sustainability. A Christian convent was chosen as the ideal site due to several strategic advantages. Its closed structure ensures a controlled environment where unauthorized or illegal power connections are not possible. This control helps to prevent system misuse and promotes operational efficiency. The clear hierarchical organization within the convent allows for the identification of responsible individuals who can oversee and manage the microgrid. This reduces the risk of corruption or misuse for personal advantage, ensuring that the system serves the convent's community. The convent's educational facilities create opportunities for direct educational outreach, enabling students and community members to engage with the system as a learning resource. The influence of the convent extends beyond its boundaries, as it holds a strong position of trust and moral authority within the local village. This influence might help the microgrid demonstrator to encourage more people of this area to consider using renewable energies, and reduce any potential mistrust there might be in relation to the use of PV power.

3.2.5.2 The Status Quo of the Designated Microgrid

The selected district for the field laboratory centres around St. Anna's convent, which is a religious and community hub. Connected to the convent are a parish, a primary school, a high school, and a hospital. The convent oversees essential community services and maintains close ties with these institutions, either by operating them or sharing the same buildings.

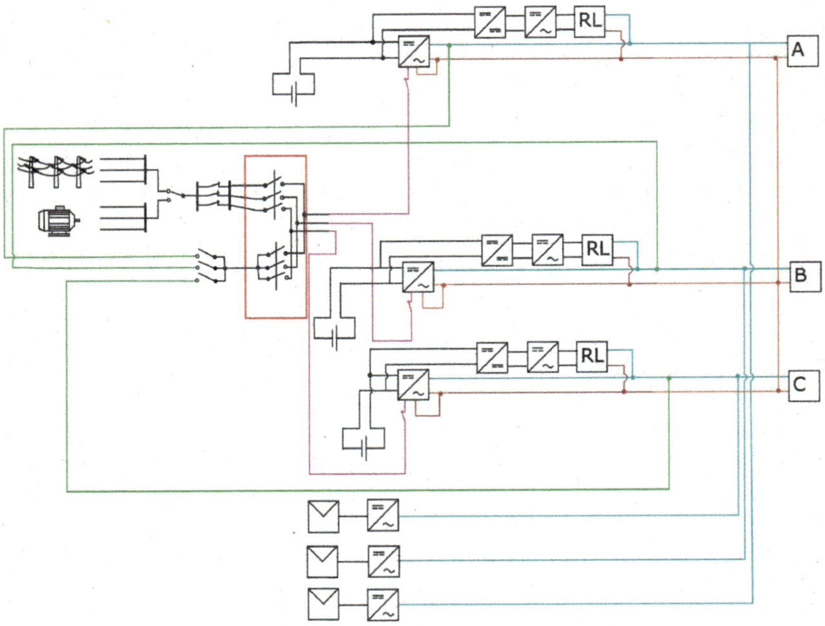

Fig. 21: Interconnection plan between the three different subsystems (A, B, and C) of the convent: .

Potential locations for the field laboratory have been the secondary school and the convent. Within the convent, the electrical infrastructure consists of three independent single-phase subgrids (see Fig. 21), with no interconnections, except for a shared three-phase water pump that is rarely active. The school is also locally divided into three independent single-phase systems with a limited number of interconnections at water-pumps and in the kitchen area.

The local public grid is unreliable, with daily blackouts requiring alternative energy solutions. A shared diesel generator supports the hospital, parish, and convent, but high fuel costs, maintenance needs, and environmental impact make this option unsustainable. Heating and cooking rely primarily on firewood and gas; electricity is only sparsely used for such energy intensive applications.

3.2.5.3 Designing Phase

To establish a rough understanding of the energy requirements of the school and the convent, an estimation of the power consumption for both premises was conducted. By analysing existing electricity and diesel fuel bills, rough load profiles on a monthly basis were developed. To further refine the understanding of energy usage patterns, a demand analysis system for continuous monitoring of energy consumption was installed by local contractors.

An initial visit to the high school and the convent was carried out as part of the preliminary assessment process. Several essential tasks were performed to support the planning, design, and implementation of the microgrid system. Since the budget of the project is limited, all electric installations and the interconnections of the different buildings could not be changed. As such, getting a precise overview of the existing electric infrastructure and the creation of detailed connection plans is crucial for the later success of the project. Precise and well-structured connection plans are essential for avoiding design errors, ensuring compatibility of system components, and facilitating future maintenance and troubleshooting. Additionally, a maximum load measurement was conducted to determine the highest level of power demand at a given moment. This information was used to define the system's capacity requirements, ensuring that the microgrid could handle peak consumption without disruptions.

To ensure the stability and quality of the energy supply, power quality measurements were performed. These measurements assessed key factors such as voltage stability, harmonic distortion, and frequency consistency. Identifying and addressing any issues related to power quality is crucial for the longevity and reliability of electrical devices connected to the microgrid.

Data validation and error correction for the previously installed measurement system were also carried out. As this system collects continuous data from the installed sensors, it is essential to ensure that this data is accurate and free from errors. Data inaccuracies can arise from sensor malfunctions, communication issues, or external disturbances. The process of validating the data and removing errors helped to improve the accuracy of energy consumption forecasts and the effectiveness of system optimization measures.

Finally, a weather station was installed at the site. This weather station provides continuous, real-time data on environmental conditions, such as solar radiation, temperature, wind speed, and humidity. This information is vital for optimizing the performance of the PV system, as it directly influences solar energy production. The weather data also serves as a reference for energy production forecasts and enables predictive maintenance of the system. Collectively, these tasks form a solid foundation for the development of a robust, data-driven, and sustainable microgrid system.

One of the key steps in designing the system was the estimation of the required capacity for PV modules, inverters, and batteries. With self-designed excel spreadsheets for system layouts as well as specialized planning tools like PVsyst, various system configurations for the school and the convent were modelled to determine the optimal capacity needed to meet the energy demands of each site. The calculations considered factors such as maximum load, daily load profiles, and weather-related fluctuations in solar energy production. The goal was to design a system that could provide stable and sufficient power supply while minimizing excess capacity and associated costs.

The following step was the estimation of hardware costs for a standard microgrid system. This included the cost of PV panels, inverters, batteries, mounting structures, control units, and supporting electrical components. In addition, shipping and customs duties were estimated since the whole grid needs to be shipped from Germany to Kenya and transported to the designated site. Accurate cost forecasting helps avoid budget overruns and ensures the project's economic viability.

Given the budgetary constraints, it was necessary to match the available components as closely as possible to the energy needs of the system. While the initial plan considered the possibility of including the school in the microgrid, it was determined that the school's energy demand was too large to be supported within the current budget and technical capacity. Additionally, the numerous connection changes required at the school added complexity to the system. As a result, the decision was made to prioritize the convent as the primary site for the microgrid. To cover the additional costs of a research system, the best way forward for all stakeholders was to implement a fuel-saver system for the convent, allowing for the integration of PV-generated electricity with the use of the existing grid connection. This approach minimizes the consumption of metered power from the grid and lowers operating costs. The diesel generator, in particular, would no longer be necessary and be replaced with a clean alternative.

The core hardware components of the microgrid include batteries, PV panels, and inverters. The battery system serves as a storage solution to provide energy during periods of low solar production or at night. Proper battery sizing and capacity planning are essential to ensure energy availability at all times. PV panels are selected and installed with a focus on maximizing energy output. In this project, the PV modules were aligned in an East/West orientation, to maximize the power output in the morning and afternoon, which are usually busier than at noon. This setup maximizes the amount of power which can be used directly from the panels, without the need of being stored in the battery system.

From a research perspective, several key goals must be considered during the implementation of the system. One of the primary objectives is to analyse how community-driven microgrids influence social and economic behaviour within the community. This includes investigating the impact of shared energy resources on community engagement, cooperative behaviour, and the equitable distribution of energy. Additionally, the research aims to evaluate the interconnection of different microgrid systems and the usage of demand-side management in a practical way. Furthermore, the system will also serve as a teaching tool, enabling hands-on learning for local students, technicians, and researchers. Data from load measurements, power quality assessments, and system performance will be used for ongoing research and optimization of future microgrid designs. By addressing these research goals, the project contributes to both scientific knowledge and local community development.

3.2.5.4 Procurement

The procurement process for the microgrid system involved several crucial steps to ensure the selection of high-quality components, reliable suppliers, and effective system integration. This process aimed to balance cost-efficiency, technical performance, and logistical feasibility.

A partner in this process was *SMA Sunbelt*, a globally recognized provider of energy solutions. This partnership ensured access to high-quality equipment under favourable conditions. The partnership with SMA Sunbelt provided technical support and access to advanced technologies, which are critical for the success of the project, while giving SMA Sunbelt the possibility to use this demonstrator grid as an example for potential customers and as a training facility for their partner companies in Eastern Africa.

Starting the whole procurement process as a university is very difficult, especially if the system cannot be tested in the laboratories. Any small error just noticed on-site in Kenya could make the whole project fail, because the procurement of suitable replacements and alternatives to get the system running would be nearly impossible in the rural areas of Kenya. To minimize the risk of malfunctions and design errors, the standard system was planned together with a specialized whole-system integrator: *Asantys* provided design, logistics, and installation support, and played a pivotal role in ensuring that the system design met the project's technical and operational goals. Their expertise in system integration allowed for efficient coordination of component selection, transportation, and on-site installation. Only the connections for the transfer of power, data or control signals to the research system were completely customized for this project.

Through this procurement strategy, the project ensured access to high-quality, compatible components and established strong partnerships with reputable suppliers and system integrators. This approach not only supported technical excellence but also reduced operational risks and enhanced the likelihood of project success.

Fig. 22: Preparation of the roof for the PV generator in Lwak after delivery of the PV modules. Foto: Stefan Krauter.

3.2.5.5 Construction and Commissioning

The construction of the system was performed by universities' employees with the support of two Asantys employees. While the engineers constructed the microgrid, the economic team finalized the contracts about ownership and later use with the sisters. Using the possibility to see all the components coming from the truck and being installed, the didactics and social science team motivated the young sisters to accompany the process in the role of junior reporters, creating a documentary about this process for others, and to gain additional insights about the role and the functionality of the individual parts and how they work together.

In Kenya, only companies or technicians with a valid Kenya Power licence are authorized to carry out electrical installations. To meet this requirement, a licensed local installation partner was engaged to oversee and execute the on-site construction work. This partnership ensured compli-

Fig. 23: Installation of the rooftop PV generator in Lwak. Foto: Stefan Krauter.

ance with national electrical safety standards and legal regulations like the governmental »no feed-in« policy back to the public grid.

The commissioning ceremony led by the local priest was a festive event for the whole convent and the surrounding village. Religious aspects like the blessing of the components with holy water might seem surprising, but effectively they increase levels of trust and enhance the population's connection to the system.

3.2.5.5.1 *Technical Research in Lwak*

The technical research associated with the microgrid system focuses on several critical aspects related to system communication, energy management, and interconnection of phases. These research activities aim to improve the system's efficiency, flexibility, and adaptability while exploring novel approaches to microgrid design and operation. While the DSM is described separately in chapter 3.3, the other technical research ideas of this grid are covered in this part.

3.2.5.5.2 *Data Acquisition and Processing*

The first part of the research involved creating a system for real-time data monitoring, storage, and processing. The system collects data from PV panels, inverters, batteries, and loads, tracking metrics like power generation, consumption, battery charge, and efficiency. Standardized protocols, Modbus and MQTT, ensure interoperability across devices. Data is stored in an InfluxDB database with a Grafana dashboard, providing real-time visualization and remote access. This enables trend analysis, anomaly detection, and performance adjustments, while remote access facilitates continuous oversight and troubleshooting without on-site visits.

3.2.5.5.3 *Interconnection of Microgrids*

One of the project's main research goals was to explore the possibilities to interconnect independent, smaller microgrids to a larger one, which could react more robustly to imbalances or power shortages in one of the individual grids. The internal structure of the convent makes it possible to emulate interconnected villages by treating each phase as a virtual village with independent energy demands, generation, and storage. A surplus of energy from one »village« can support others, enabling decentralized, cooperative microgrid behaviour. A controlling Raspberry Pi analyses the current power generation, usage, and grid stability of each »village«. If one of those villages struggles but others have a healthy power supply, they can interconnect automatically and support each other with a potential power surplus. Using an agile master-slave hierarchy, the weaker phases automatically couple directly to the AC-site of the master subgrid. This method eliminates the need for an expensive DC link, where power must be repeatedly converted and transformed between phases. As a limitation, the phase-interconnection is feasible only in battery-powered mode due to the phase-angle synchronization requirements, which are not possible with a connection to the public grid.

3.2.5.5.4 *Overall Aim*

The overall technical research aims to develop a robust, data-driven, and adaptive microgrid system. By improving data monitoring, storage, and processing, the system can achieve real-time decision-making and predictive control. Interconnection of phases allows for greater system flexibility, while autonomous coupling and decoupling increases system resilience. The emulation of interconnected village microgrids provides valuable insights into cooperative energy sharing models. Together, these research activities aim to create a scalable, adaptable, and community-focused energy system that can be replicated in other regions with similar energy needs.

3.2.5.5.5 *Demonstration Purposes/Multiplicator*

The microgrid system serves as a powerful demonstration model with a strong multiplier effect, providing educational, technical, and social benefits to various target groups. Its impact extends beyond the immediate operational goals, promoting knowledge transfer, community empowerment, and long-term sustainability.

Young Sisters: The system acts as a living laboratory for young sisters, teaching renewable energy, energy efficiency, and sustainability. By engaging with components like PV panels and inverters, the sisters acquire technical skills and energy literacy for future leadership roles.

Secondary School: The microgrid enhances STEM education by providing hands-on experience with renewable energy technologies. Students learn about energy production, storage, and management, inspiring interest in green energy careers and fostering data-driven decision-making skills.

Village Population: For the village population, the system acts as a role model, showing how efficient, clean, renewable energy can be used. The fact that even the highest local religious authorities support this form of energy increases this effect and might lead local consumers or business owners to consider using a PV system.

Cooperation with SMA Sunbelt: Through the cooperation with SMA Sunbelt, the grid can be used to train local electricians or installers, and it can help build up local capacities, which is an important step for a sustainable clean energy system.

3.2.5.6 Lessons (to be) Learned

During large projects, various smaller or larger issues can arise, potentially reducing sustainability or, in the worst case, leading to a complete failure. However, many of these potential problems can be addressed in advance through proactive planning and effective risk management to prevent them or minimize their impact.

Innovative projects are often planned and coordinated by individuals, and yet these people, whether intentionally or unexpectedly, may leave their positions. This can result in a lack of contact persons or insufficient accountability from successors. In such cases, it can be helpful to assign responsibilities not only morally to specific individuals but also, if necessary, contractually to specific roles or positions.

Without skilled professionals on site, such a project cannot be implemented. It is therefore crucial to have not only reliable decision-makers locally but also electricians to accompany the development process and carry out future maintenance and repair tasks. Since younger employees in particular often change positions at short notice, it is essential to establish backup personnel and local knowledge clusters at the outset to mitigate the impact of staff turnover.

Amid all the enthusiasm for planning and implementing such projects, the focus often remains solely on the initial costs. Equally important, however, is the need to clarify responsibilities and potential funding for maintenance and operational costs.

For the importation of components, experienced, reliable, and trustworthy dispatchers with a long track record in the area should be contracted to avoid long delays and unexpected expenses.

3.3 Demand-Side Management

Josephine Nakato Kakande, Stefan Krauter, Godiana Philipo

3.3.1 Introduction

Demand-side management (DSM) refers to the influencing of customer electricity usage (e.g., by indication lights, tariffs, or load shedding) so as to change the pattern of the demand profile of a customer or group of customers (Electric Power Research Institute (EPRI), 1993). By synchronizing electricity supply with demand, problems such as system overloading, blackouts, and accelerated equipment deterioration as well as excessive costs for energy storage can be alleviated. In microgrids, which lack the generation capacity and system inertia associated with the grid, particularly when operating in isolated mode, DSM techniques are even more essential.

DSM measures can be put into effect by the energy supply company or by the electricity end-users. In general, energy efficiency and demand response measures such as incentives or time-of-use (TOU) tariffs are used to alter consumption patterns in order to promote sustainable development and conservative, more efficient use of energy resources (Bakare et al., 2023; Nebey, 2024).

Measures that have been taken to promote energy efficiency in Africa, including East Africa, include prepayment meter roll-out, promotion of compact fluorescent lamps (CFLs) or light emitting diode (LED) lamps, sensitization and creating awareness, power factor correction, and reduction in grid losses (CPCS, 2021). Distributed generation and energy storage have also been recommended as DSM strategies (Nebey, 2024). By establishing energy generation systems closer to the loads, microgrids can reduce grid losses or, in the case of stand-alone systems, avoid them altogether.

DSM is a key area of interest in East Africa. The Uganda Energy Policy 2023 spells out DSM strategies such as the development of standards and regulations for energy efficiency in households, institutions, industry, and commercial sectors, promotion of clean cooking technologies, implementation of energy management systems in high-demand institutions, as well as promotion of energy-efficient and green cooling, agriculture, and transportation. The policy guide also highlights the need for increasing renewable energy supply such as through off-grid solar PV systems.

In Tanzania, several DSM options were recommended, such as domestic refrigerator replacement by more efficient appliances, CFL disbursement, direct load control by the utility TANESCO (Tanzania Electric Supply Company Limited) to switch off participating commercial customers' air conditioners when the grid is stressed, TOU tariffs and incentives (CPCS, 2021).

A study carried out in Kenya found that while DSM awareness was low, household uptake of renewable energy, price responsive demand, implementation of energy efficiency and conservation measures, and integration of smart appliances were the most viable options in Kenya (Kiprop et al., 2020).

Energy efficiency measures are easier to roll out in developing economies, e.g., for lighting, while demand response may require more advanced, smarter technologies (Islam et al., 2022). However, grid-edge, the Internet of Things, and smart technologies are becoming more and more ubiquitous, and Africa, including East Africa, is better placed than ever to adopt them to enhance its energy sustainability.

3.3.2 Typical DSM Approaches

Common DSM techniques include the following (Meliani et al., 2021):

- Load shifting whereby demand is shifted from peak to off-peak hours by running shiftable or deferrable loads during off-peak hours
- Peak shaving in which some loads are curtailed to lower demand during peak hours, in order to lower maximum demand
- Valley filling whereby the demand during off-peak hours is raised, for instance by introducing new loads like electric vehicles
- Strategic load growth in which measures are undertaken to achieve longer term increase in demand
- Strategic conservation that attempts to lower overall demand, for instance, by implementing energy efficient measures

- Flexible load shape whereby customers can modify their demand profile in tandem with agreements with the energy supplier.

These six approaches are depicted in Fig. 24 (Lampropoulos, 2014).

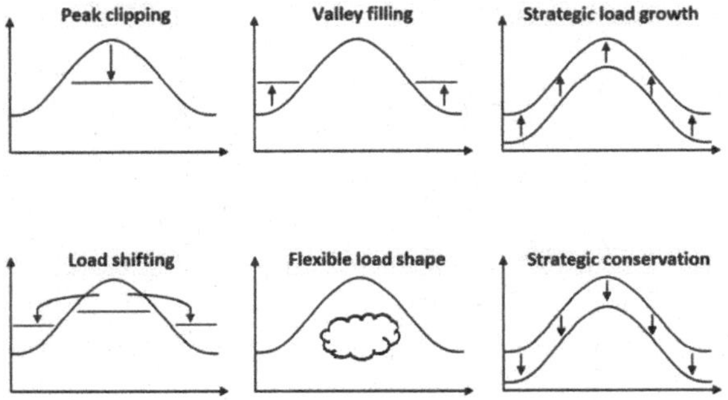

Fig. 24: Demand-side management techniques (adapted from Lampropoulos, 2014).

3.3.3 DSM Case Study in Lwak, Kenya

As part of the A:RT-D Grids project, an experimental study was carried out in Lwak, Eastern Kenya. The refrigeration appliances at the site were selected for DSM measures. Refrigeration is typically the largest energy consumption load in residential settings, as was found to be the case in Lwak Convent in Kenya (Kakande et al., 2023).

The IEA predicts a significant increase in refrigeration and cooling loads in Africa (International Energy Agency (IEA), 2023). Our focus on DSM in refrigeration aligns with the UN's Sustainable Development Goals (SDG) 2030 targets of SDG 2 on zero hunger and SDG 7 on sustainable, clean energy for all (United Nations, n.d.).

3.3.3.1 Site Description, Methodology, and Findings

The site has 8 refrigeration appliances and is characterized by separate buildings containing the electrical equipment and loads (see Fig. 25).

Fig. 25: Site layout in Lwak, Kenya (Kakande et al., 2025).

The platform uses IOT tools, including smart plugs, cloud services, and Raspberry Pi controllers. Open-source solutions were used to set up the monitoring and control platform. A docker platform with Node-RED, MQTT, and InfluxDB was configured on the Raspberry Pis. Because the solar system and the load controller were using different wireless networks, a cloud-based HIVE MQ broker was used to relay battery state of charge information from the Sunny Island inverters using a Raspberry Pi connected to the PV system network to the main Raspberry PI microcontroller located at the load centre. MQTT, Docker, Influx DB, and Grafana were the messaging, data storage, and data visualization tools used, with internet connectivity, WiFi, and Ethernet connections used for communication (see Fig. 26).

Shelly Plus 1 PM devices were used to control the on and off state of the refrigeration appliances based on SoC information.

Using a similar concept to pre-cooling, different temperature thresholds were set and used to determine when the load controller would switch the refrigeration appliances on and off, depending upon the battery SoC.

Variable thresholds were used for the control of appliances. For Building 1, T_{low1} was set to -14°C and -15°C, T_{high1} (°C) to -12 °C, T_{low2} (°C) to -10 °C and T_{high2} to -6 °C, to cater for varying baseline internal temperatures of the freezers. The battery threshold SoC was set to 75 %, above which the appliance would operate at lower temperatures, and below which the appliance would operate at higher temperatures, thus modifying the energy draw by the appliances. SoC values for Building 3 were used in the analysis.

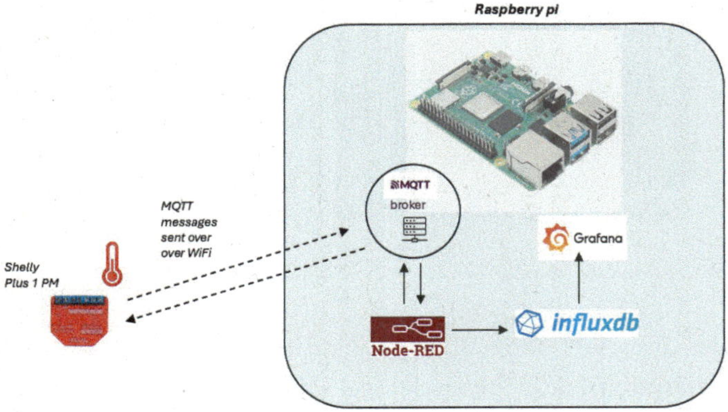

Fig. 26: Communication tools to carry out DSM in Lwak (Kakande et al., 2025).

The average daily power values for Freezer 1 appliances are given in Tab. 2. Average power reduction is greatest for Freezer 1 at 30 %, followed by Fridge 2 at 13 %, while Freezer 3's energy gains are negligible (see Fig. 27), which corresponds to the respective percentage average daily energy savings with DSM for those appliances. Ambient temperatures and user interactions are not considered in the analysis (Kakande et al., 2025).

Tab. 2: Average daily refrigeration power draw with and without DSM (Kakande et al., 2025)

Appliance	No DSM	With DSM	
	Average power (W)	Average power (W)	Average power change relative to Case 1 (%)
Freezer 1	97.6	68.4	-30 %
Fridge 2	137.1	119.3	-13 %
Freezer 3	114.2	111.5	-2 %

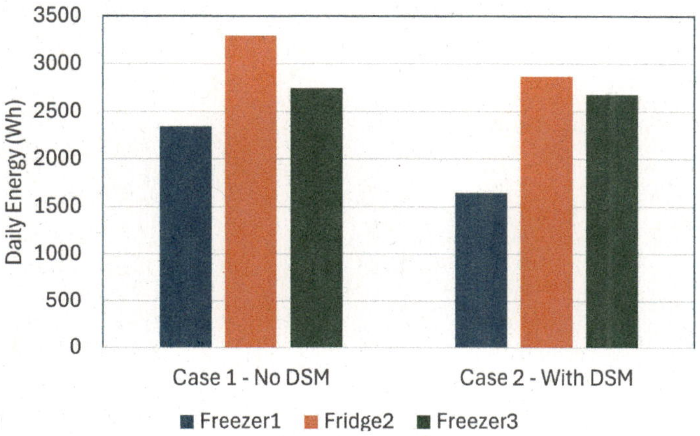

Fig. 27: Energy drawn by appliances with and without DSM (Kakande et al., 2025).

Energy consumption was reduced by adjusting the refrigeration appliance temperatures based on battery SoC. Specifically, appliances are cooled to lower temperatures when the battery SoC is high, whereas higher temperature thresholds are employed when the battery SoC is low, thereby reducing the overall cooling energy demand.

Further analysis of Building 2 was done based on different temperature thresholds, and pre-analysis and monitoring of the fridge and freezer temperature was done to observe their minimum and maximum operation. Then the temperature thresholds were set accordingly to avoid food spoilage. According to literature, the safe recommended fridge temperature is 1.1°C to 4.4°C (Kakande et al., 2023), and 2°C to 4°C is recommended for ready-to-eat foods. Other sources reported 5°C to 8°C as refrigerator temperature thresholds Postnikov et al., (2019). For freezers, Postnikov et al., (2019) have the recommended temperature as below -12°C, while Tran et al., (2015) state that the temperature should be below -18°C.

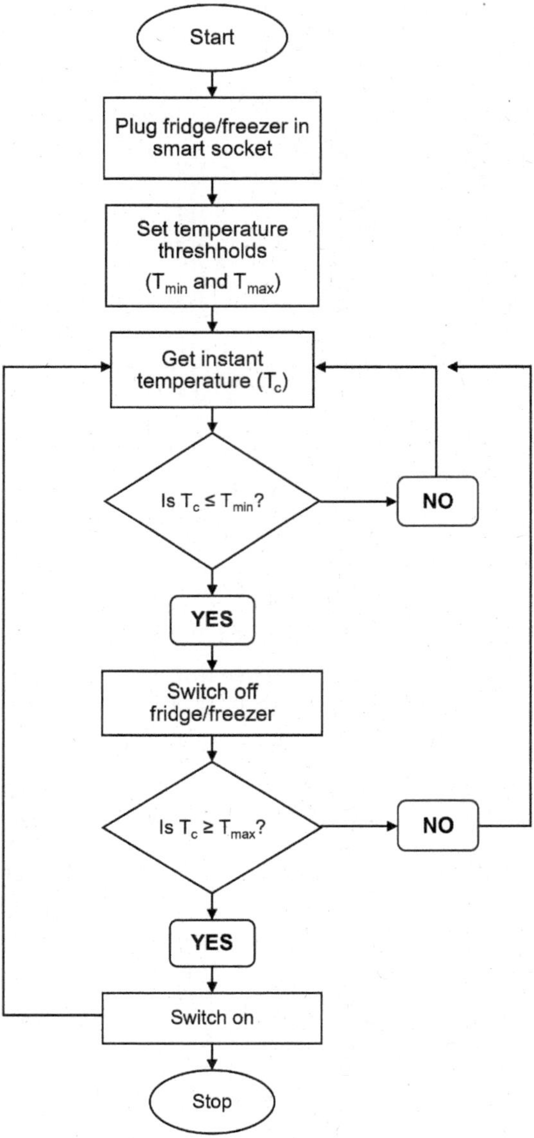

Fig. 28: Flow chart of the cooling appliances' control mechanism of switching ON and OFF appliances based on temperature.

It is worth noting that placing a fridge/freezer in a cooler environment may result in lower energy consumption since consumption may increase with a higher external temperature. Also, other studies recommend that consumers should not place their refrigeration units in areas of temperatures below 55°F (12.8°C) or above 90°F (32.2°C) (Burgess, 2015).

Fig. 28 shows a flow chart of suggested load control usage where the instant temperatures are monitored and compared with the thresholds set to determine the appliance's on and off times.

Results show that some of the refrigeration appliances, e.g., Freezer 8 represented by profile in Fig. 29, demonstrated less energy-consuming profiles after load control, meaning that we can save energy by switching their operation time while preserving food as before.

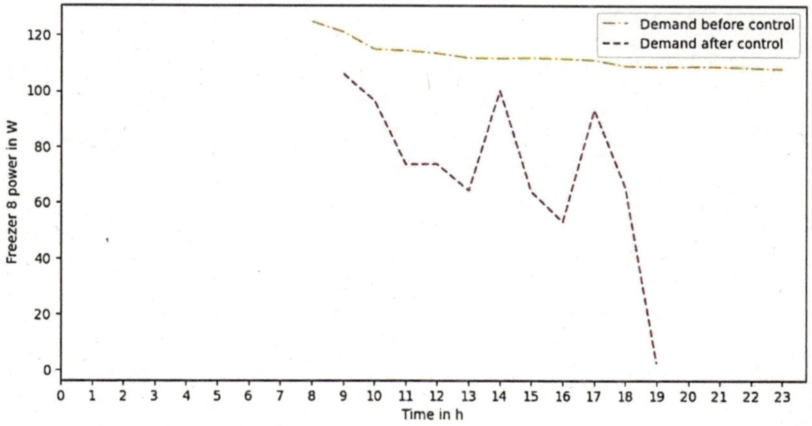

Fig. 29: Daily load profile before and after temperature-based control for Freezer 8.

Worse results were observed for other refrigeration units such as Fridge 4 after temperature-based control (its profile is depicted by Fig. 30) whereby the unit tends to consume more energy instead of saving it. In addition, it was reported that the unit had poor cooling capability and could not adequately store items (food) as before. The explanation as to why poor results were obtained was thought to be due to their poor insulation and inefficiencies which made them unable to maintain the temperatures, consuming more energy with more temperature fluctuations. The results pointed to the challenge of finding the right type of control and suggested improved temperature-based control methods for inefficient fridges or freezers since they suffer from limited temperature stability, which might compromise food safety and quality.

However, the trade-off of a load-shift that would result in higher consumption could even be positive, if the cost advantage caused by the load-shift (e.g., cheaper electricity during consumption periods, reduced costs for storage) is higher than the expense of the increased consumption by the load-shift.

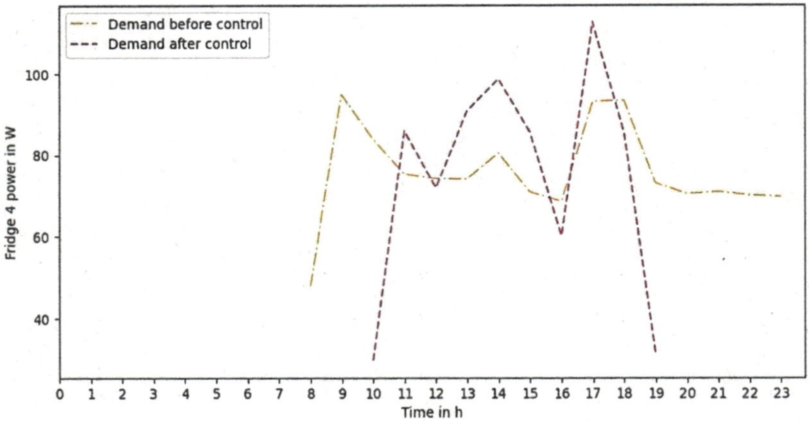

Fig. 30: Daily load profile before and after temperature-based control for Fridge 4.

3.3.4 Conclusion and Outlook

DSM was applied to refrigeration appliances using battery SoC values to lower consumption during low to no PV production and low battery SoC periods by applying temperature-based control. These results promised more energy-saving potential by switching off refrigeration appliances which should be done without compromising the food stored. However, the load-shifting potential in Lwak turned out to be relatively low due to existing conditions (low insulation/inefficiencies of some refrigeration units).

Therefore, it is crucial to assess and recognize the limitations of inefficient appliances before using temperature control to sustainably exploit the benefits of temperature-based control. Replacing new freezers with modern insulation properties can expect a significantly higher time shift potential. As an additional advantage, the total energy demand would be reduced.

Furthermore, Phase Change Materials can enable a larger time shift (with its resulting advantages) and a reduced frequency of switching the refrigerators' motors and compressors on and off, which could extend the lifetime of the equipment.

3.3.5 References

Bakare, M. S., Abdulkarim, A., Zeeshan, M., & Shuaibu, A. N. (2023). A comprehensive overview on demand side energy management towards smart grids: challenges, solutions, and future direction. *Energy Informatics, 6*. https://doi.org/10.1186/s42162-023-00262-7

Burgess, T., S. *The* Effects *of External Temperature on the Energy Consumption of Household Refrigerator-Freezers and Freezers*. Master Thesis at Texas A& M University, TX, USA, August 2015.

CPCS. (2021). *Draft Report on Guidelines for Demand-Side Management*. https://rerasadc.com/wp-content/uploads/2021/09/DSM-Guideline-Report.pdf

Electric Power Research Institute (EPRI). (1993). *Principles and Practice of Demand Side Management*. https://www.epri.com/research/products/tr-102556

International Energy Agency (IEA). (2023). *Africa Energy Outlook 2022*. https://doi.org/10.1787/g2120ab250-en

Islam, M. A., Hossain, M. S., Hasan, M. R., Babu, M. S. I., Banna, H. Al, Rashed, M. N. H., & Alam, M. S. (2022). Energy conservation and load shifting based DSM approach in the residential sector: applicability and impacts analysis in the case of developing countries. *2022 International Conference on Innovations in Science, Engineering and Technology, ICISET 2022*, February, 498–503. https://doi.org/10.1109/ICISET54810.2022.9775910

Kakande, J. N., Philipo, G. H., & Krauter, S. (2023). Demand side management potential of refrigeration appliances. *2023 IEEE PES/IAS PowerAfrica*. https://doi.org/10.1109/PowerAfrica57932.2023.10363161

Kakande, J. N., Philipo, G. H., & Krauter, S. (2025). Smart demand side management using battery SoC-based control for refrigeration in a solar hybrid system. *IEEE Global Power, Energy and Communication Conference 2025* (IEEE GPECOM 2025).

Kiprop, E., Matsui, K., & Maundu, N. (2020). The future contribution of demand side management to solving Kenya's energy insecurity problems. *International Journal of Environmental Science and Development, 11*(3), 111–115. https://doi.org/10.18178/IJESD.2020.11.3.1235

Lampropoulos, I. (2014). *Energy Management of Distributed Resources in Power Systems Operations.* PhD Thesis. Eindhoven: Technische Universiteit Eindhoven. doi:10.6100/IR771935

Meliani, M., Barkany, A. El, Abbassi, I. El, Darcherif, A. M., & Mahmoudi, M. (2021). Energy management in the smart grid: state-of-the-art and future trends. *International Journal of Engineering Business Management, 13,* 1–26. https://doi.org/10.1177/18479790211032920

Nebey, A. H. (2024). Recent advancement in demand side energy management system for optimal energy utilization. *Energy Reports, 11,* 5422–5435. https://doi.org/10.1016/j.egyr.2024.05.028

Postnikov, A., Albayati, I.M., Pearson S. et. al. (2019) Facilitating static firm frequency response with aggregated networks of commercial food refrigeration systems. *Applied Energy,* Volume 251, 113357. https://doi.org/10.1016/j.apenergy.2019.113357

Tran, J. Gilles. J., Mann, R., Murthy, V. *Automated Demand Response Refrigerator Project.* Project Report, University of Berkely, CA, USA, October 2015.

United Nations. (n.d.). *The 17 SDG Goals.* https://sdgs.un.org/goals

3.4 Towards Sustainable Microgrids Development: Role of Wireless Sensor Systems and Signal Processing—Case of a Rural Community Microgrid in Tanzania

Ibrahim Mwammenywa, Ulrich Hilleringmann

3.4.1 Introduction

Building upon the discussions presented earlier, the proliferation of microgrids is primarily enabled by the significant integration of renewable energy sources (RES), including wind, solar, biomass, and hydro. Microgrids play a key role in providing affordable, reliable, and sustainable energy, particularly in rural regions, in line with the United Nations Sustainable Development Goal (SDG) 7 (UNGA, 2015; PwC, 2016). Despite their potential, many microgrids are often affected by unreliability and recurrent blackouts due to the inherent intermittency and stochastic nature of RES (Hilleringmann et al., 2021). The absence of effective real-time monitoring and control strategies exacerbates these challenges, resulting in frequent power imbalances and disruptions (Hilleringmann et al., 2021).

Recent advancements in signal processing and communication technologies, including the Internet of Things (IoT) and Wireless Sensor Networks (WSN), have significantly influenced electrical grid operations. These technologies have been transforming microgrids from the classical and conventional microgrids of only a power network into an advanced microgrid of both power and communication networks. The transformation of microgrids has been largely necessitated by the need to implement monitoring and control strategies that rely on bidirectional information flow among grid components. The employment of proper communication infrastructures in microgrids helps in controlling and managing the reliability and sustainability of microgrids despite the intermittent nature of RES.

The intelligence of a microgrid is largely determined by the capability of its communication infrastructure to transmit real-time data between system nodes. Smart microgrids handle large volumes of data and critical operations, requiring reliable, ubiquitous, and low-latency communication. Therefore, selecting a communication technology that ensures adequate reliability, security, and performance—considering parameters such as bandwidth, latency, and packet loss—is crucial for microgrid sustainability (Serban et al., 2020). The communication system is a fundamental component of smart grids, including smart microgrids, and requires coordination and integration of all heterogeneous elements. To support interoperable communication standards, the IEEE provides guidelines and standards for communication in electric grids (Bush et al., 2013).

As mentioned above, the objective of communication in a microgrid is to make the microgrid stable, robust, reliable, and sustainable through different technological applications. Technological applications, such as Advanced Metering Interface (AMI) and smart meters, home appliances, and systems such as Demand-Side Management (DSM) and Demand Response (DR) intensely require real-time supervision of the microgrid. Communication provides the backbone for all these applications in the microgrid. Fig. 31 shows how communication plays an important role in attaining the sustainability of microgrids.

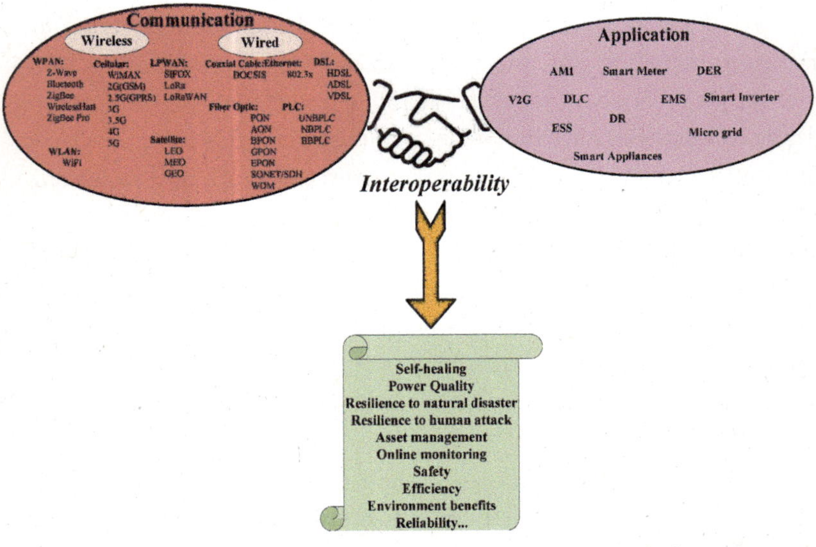

Fig. 31: Role of communication in attaining a sustainable microgrid (Tightiz & Yang, 2020).

3.4.1 Communication Media in Electric Grids

The elements of a microgrid, such as power generators (e.g., PV panels, wind, and hydro turbines/generators), storage facilities (such as rechargeable batteries systems), loads, and other devices (such as charge controllers, power electronic converters, and inverters), communicate among each other in the form of data packets. For the information to move from one element to another, there must be a communication network. The communication network can be of a wired, wireless, or hybrid technology, as summarized in Tab. 3.

Signals inside elements, such as power electronic converters, controllers, and measurement devices, are normally transferred through wired media over a short distance. However, for signals from more distant elements such as smart meters with different users, fibre optics, power line communication (PLC), or wireless communication media are recommended. The versatility and scalability of wireless technology and low cost of maintenance offer better economic value when compared to wired media which requires a high cost of implementation and maintenance. Moreover, PLC has some technical shortcomings which limit the scale of the network, as data frames cannot be propagated through transformers in the electric network (Wang et al., 2011).

Tab. 3: Comparison of communication technologies in the context of microgrids (Serban et al., 2020)

Media Type	Technology	Maximum Coverage	Data Rate	Frequency Band	Advantages	Disadvantage
Wired	Optic Fibre	100 km	10 Mbps–100 Gbps	Near IR	High speed, Low interference, Low latency	Expensive, Complex scalability
	Ethernet (twisted pairs)	100 m	10 Mbps 10 Gbps	< 25 MHz <500 MHz	Easy access, Easy configuration	Costly, Complex scalability
	Power line	200 m	<200 Mbps	< 100 MHz	Low cost	Inflexible, Complex scalability
Wireless	Wi-Fi	< 100 m	11 Mbps–10 Gbps	2.4/5.9 GHz	Cheap, Easy configuration, Easy scalable	High security need, Susceptible to interference
	ZigBee	75–100 m	2 Kbps	GHz	Cheaper, Low power usage	Very low speed
	Bluetooth	10 m	1 Mbps	GHz	Cheaper, Low power usage	Low penetration, High security need
	Cellular 3G	50 km	< 2Mbps	800/1800/ 2100 MHz	Requires ISP	Costly
	Cellular 4G	50 km	< 10Mbps	2600 MHz	Requires ISP	Costly
	Cellular 5G	50 km	< 10Gbps	Sub-6 GHz	Requires ISP, Low Latency	Costly, Not available everywhere
	LoRa	< 10 km	< 50kbps	433/868/ 915 MHz	Cheaper, Low power usage, Easy scalable, Easy penetration	Low speed

3.4.3 Implementation of Load Monitoring and Control Using LoRa and Fuzzy Logic in a Community Owned Microgrid

In this study, the selected community is situated in a rural area of Tanzania and is served by a community-owned solar PV microgrid. Centrally located within the village, the system supplies approximately 60 households and shops spread across a 1 km radius from the plant. The microgrid is rated at 15 kWp and integrates a battery energy storage system comprising 24 lead-acid dry cells, each with a capacity of 2 V DC and 1236 Ah. As depicted in Fig. 32, the microgrid consists of three single-phase distribution lines, two of which supply households exclusively, while the third delivers power to both households and shops in the village center.

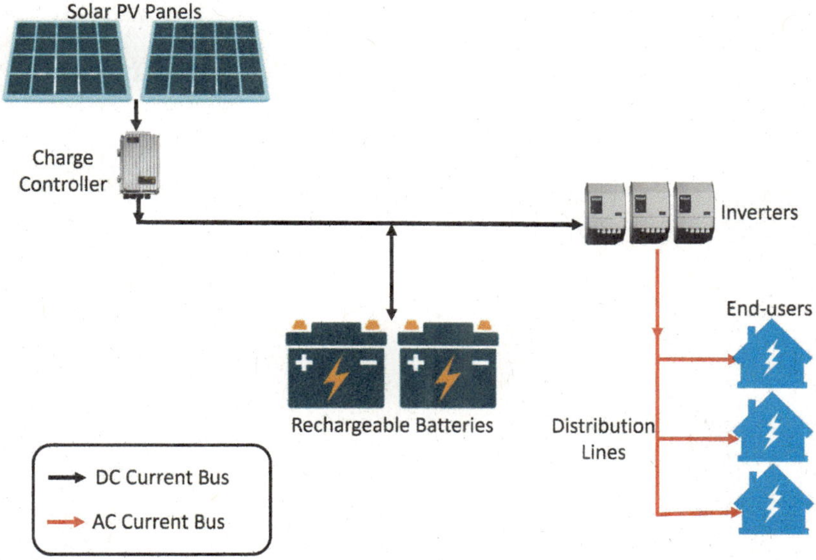

Fig. 32: Community microgrid schematic diagram of case study (Mwammenywa, 2024).

The microgrid is critically affected by overloading, primarily due to mismatches between PV generation and load demand. The absence of coordination between actual power output and overall demand has resulted in frequent power outages and, consequently, an unreliable supply to the village. To mitigate this challenge, a wireless sensor network (WSN) utilizing

LoRa technology was designed. This system enables real-time monitoring of consumer demand and regulates load consumption in accordance with PV generation levels and available battery storage (Mwammenywa, 2024). The conceptual framework and operating principle of the LoRa-based system are presented below.

3.4.3.1 System Architecture and Design

To achieve real-time load monitoring and control within the microgrid, a wireless sensor system was implemented, as depicted in Fig. 33. Consumer modules were deployed at end-user points to enable load monitoring, control, and tariff administration. The host computer is responsible for managing demand in relation to the available supply, considering both PV-generated power and stored battery energy.

Fig. 33: Decentralized load monitoring and control architecture in the microgrid (Mwammenywa, 2024).

Fig. 33 depicts a system in which consumer modules communicate with the host computer through a LoRa wireless network. LoRa, a Low Power Wide Area Network (LPWAN) technology, facilitates this communication. The host computer acquires load data from all consumer modules and compares it with real-time PV generation and the available battery storage capacity (Mwammenywa, 2024). Subsequently, it transmits control signals to the consumer modules, specifying load limits required to maintain the long-term stability of the microgrid.

3.4.3.2 Consumer Module

The consumer module, illustrated in Fig. 34, is structured into five distinct units: the power supply unit, sensing unit, processing unit, communication unit, and display unit.

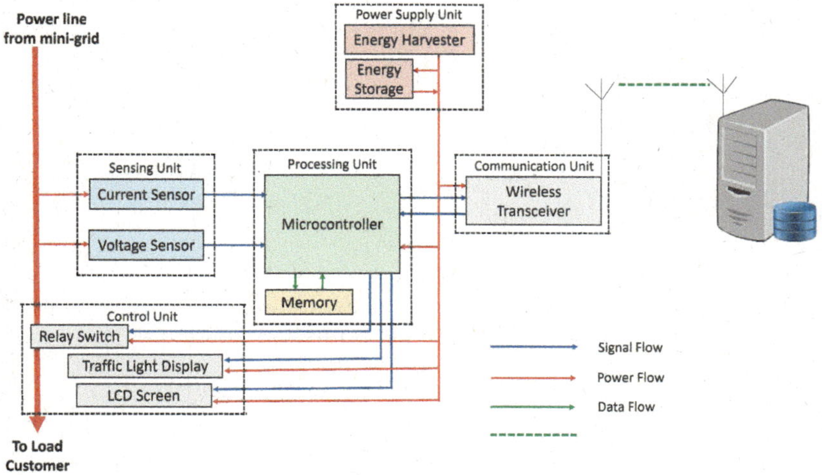

Fig. 34: Scheme of power line monitoring and supply control for customers (Mwammenywa, 2024).

Sensing Unit: The sensing unit is designed to continuously monitor the load of each consumer by measuring real-time voltage and current. The electrical power consumed by the load is computed as a function of the instantaneous current and the voltage applied across the load.

Processing Unit: An Arduino 8-bit ATmega328p micro-controller was employed as a processing unit in the consumer module. The ATmega328p micro-controller provides various functions and development options, and it was designed to carry out all logical functions and decisions in the consumer module, while providing the bus connection to all other units.

Control Unit: The main objective of the system is to manage the demand side to ensure that power consumption does not exceed available generation, thereby promoting microgrid sustainability and preventing overloading. To accomplish this goal, the system advises consumers to adjust their power usage based on the total available energy. For this study, two hierarchical levels of control have been implemented for DSM, as depicted in Fig. 35.

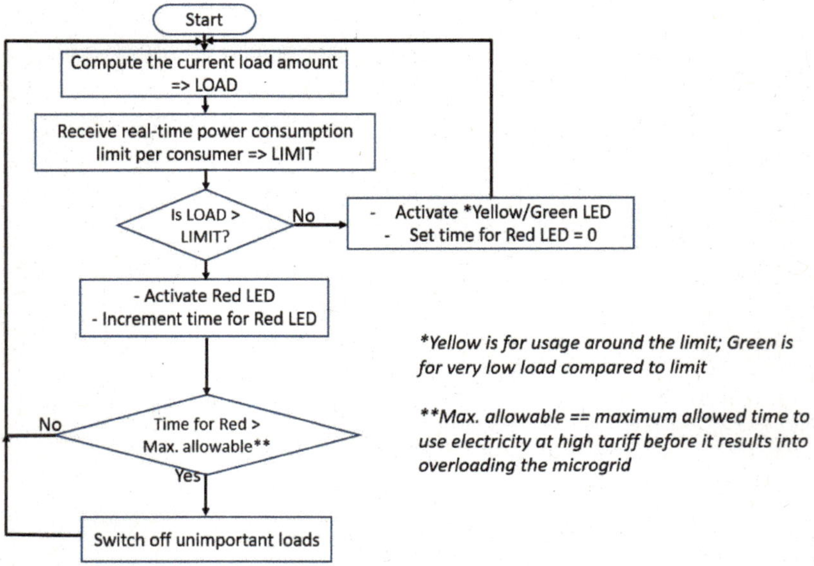

Fig. 35: Flow chart for demand response and load management (Mwammenywa, 2024).

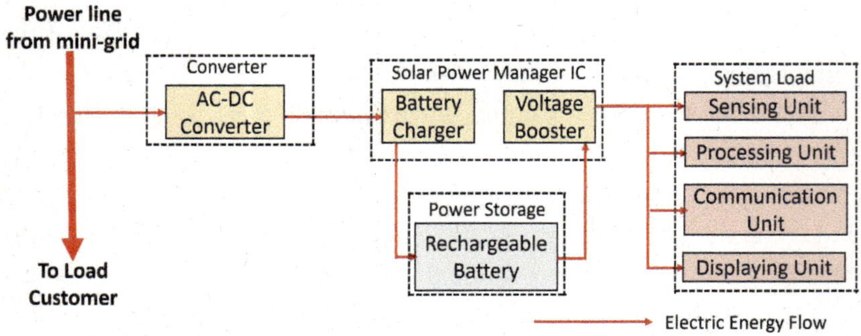

Fig. 36: Energy harvesting from AC power line (Mwammenywa, 2024).

Power Supply Unit: The consumer modules have been designed with a standalone, stable, and dependable power supply which comes from the microgrid's power lines and has integrated rechargeable batteries. This configuration ensures continuous operation of the modules during microgrid outages or faults. Fig. 36 depicts the block diagram of the power line energy harvester, which incorporates an AC-DC converter supplying 5 V DC to power the prototype system.

3.4.3.3 Wireless Communication Unit

Data communication between the consumer module and the host computer occurs via a LoRa network. The LoRa modules used in this study were designed and developed at Paderborn University, providing a transmission range of approximately 2 km, which aligns with typical microgrid coverage in SSA (Mwammenywa, 2024). The system operates on the 868 MHz frequency band, the ISM band authorized for Africa. The transceivers, based on RFM96 chips, offer high immunity to signal interference and include a matching network and clock source, simplifying transceiver construction as shown in Fig. 37 (Mwammenywa et al., 2022).

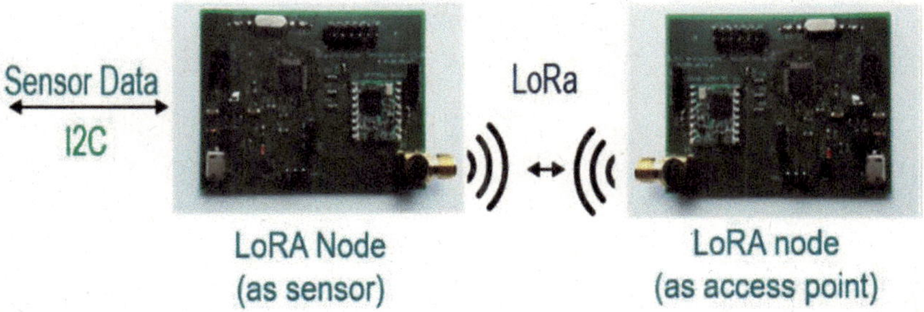

Fig. 37: Bidirectional data communication via LoRa transceivers (Mwammenywa et al., 2022).

3.4.3.4 Host Computer

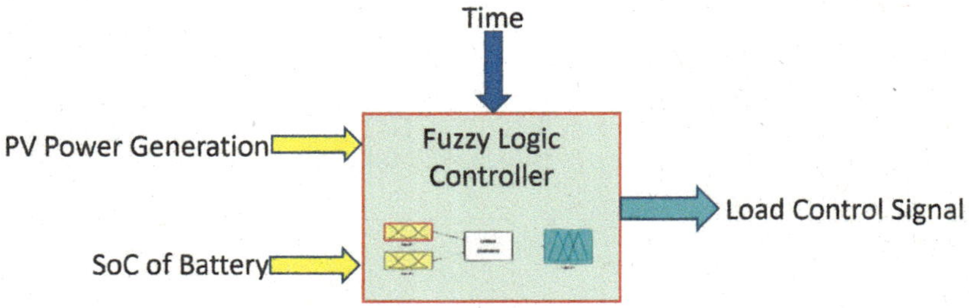

Fig. 38: FIS-Based Model for Load Control Signal (Mwammenywa, 2024).

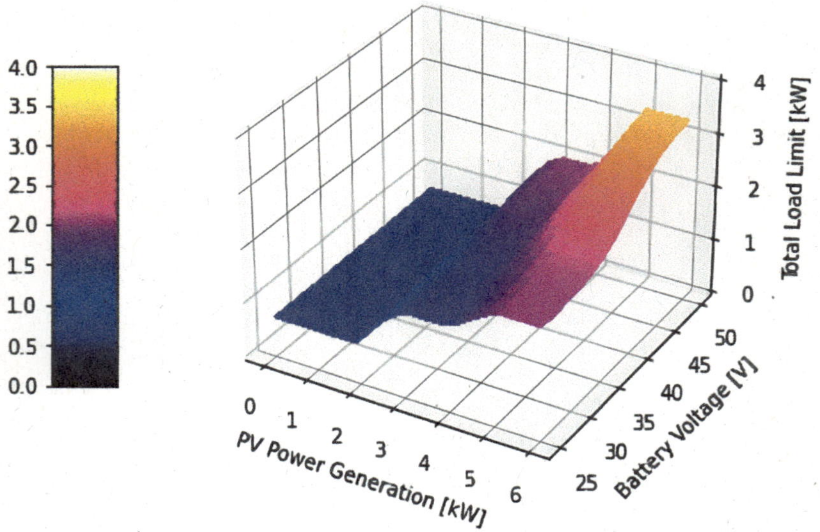

Fig. 39: Correlation Between Inputs (PV Generation and Battery SoC) and Output (Total Load Limit) (Mwammenywa, 2024).

In this design, the host computer receives real-time data on power consumption from the consumer modules, PV generation, and battery State of Charge (SoC). The data are processed by the host computer, which generates power control signals that are wirelessly transmitted to the consumer modules (Mwammenywa, 2024). As illustrated in Fig. 33, each consumer module communicates with the host computer via LoRa technology. A fuzzy logic–based controller is implemented within the host computer application. The controller generates load control signals based on real-time PV generation and battery SoC, as shown in Fig. 38. For this study, the load control Fuzzy Inference System (FIS) is modeled using a Takagi-Sugeno approach, summarized in Fig. 39.

3.4.3.5 Deployment

The implemented devices, such as consumer modules and LoRa transceivers, were installed in each house or business connected to the microgrid. Fig. 40 shows the consumer modules at the user's house communicating to the host computer at the microgrid equipment house.

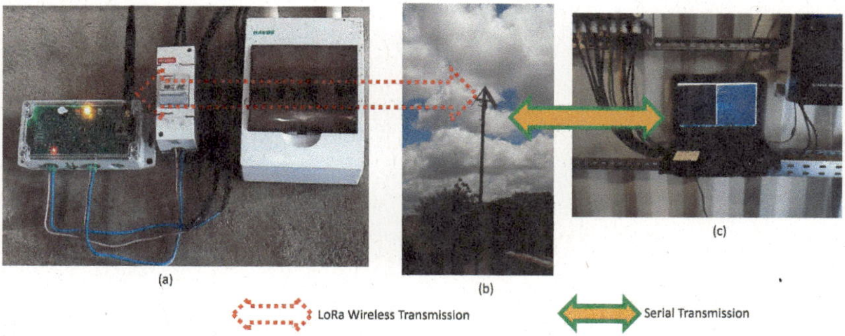

Fig. 40: Flow of Information from Installed Module Through LoRa Transceiver to Host Computer (Mwammenywa, 2024)

With these devices added to this Microgrid, its schematic shown in Fig. 32 is improved to a new schematic shown in Fig. 41.

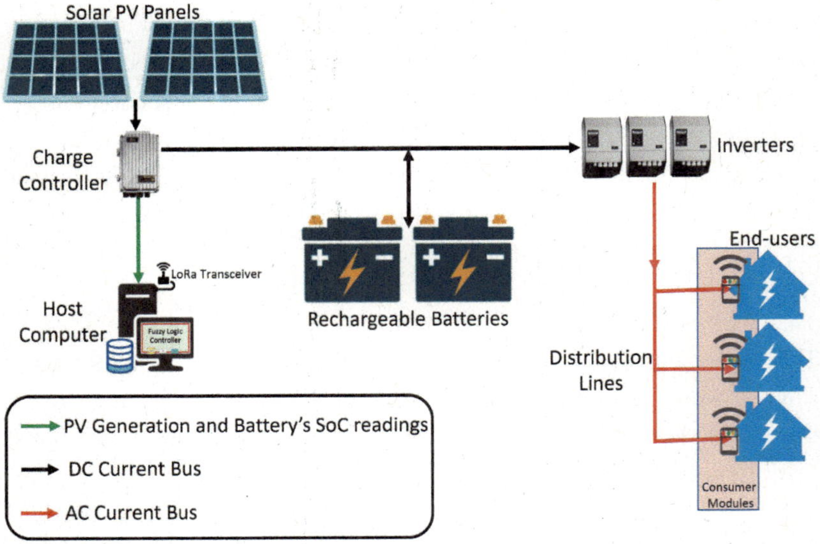

Fig. 41: Updated Schematic Diagram of the Community Microgrid (Mwammenywa, 2024).

3.4.3.6 Case Study Observation

The consumer modules were receiving signals from the host computer which could make a respective traffic light based on the loading of that particular consumer, as shown in Fig. 42.

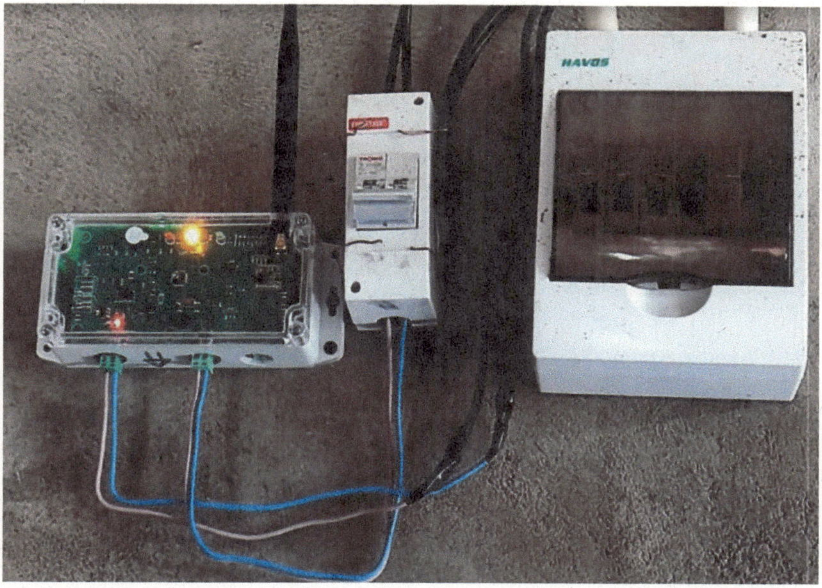

Fig. 42: Example for indication by »traffic light«: Yellow light signalling to alert a consumer of usage being close to a load limiting threshold (Mwammenywa, 2024).

3.4.4 Conclusion

Wireless communication and signal processing can play important roles in the sustainability of microgrids. The wireless sensory system has turned an unreliable community-owned microgrid into a reliable microgrid. The sensory system presented above employed the LoRa wireless communication technology, but other options are also viable.

Monitoring and control systems are essential for ensuring that microgrids operate efficiently and reliably. These systems not only reduce the occurrence of unplanned power outages but also facilitate the identification and resolution of technical faults. The deployment of a LoRa-based load monitoring and control system in this community microgrid has enhanced operational reliability, providing a stable and available power supply and contributing to the overall sustainability and robustness of the microgrid.

Moreover, a communication system in microgrids improves the customer experience and can even be used in shaping consumer behaviour. With a communication system in place, information can be bidirectionally shared between customers and microgrid operators in real time as achieved in the microgrid.

3.4.5 References

Bush, S. F., Goel, S., & Simard, G. (2013). IEEE vision for smart grid communications: 2030 and beyond roadmap. In *Proceedings of IEEE Vision for Smart Grid Communications, Beyond Roadmap* (pp. 1–19). IEEE. https://doi.org/10.1109/IEEESTD.2013.6690098

Hilleringmann, U., Petrov, D., Mwammenywa, I., & Kagarura, G. M. (2021). Local power control using wireless sensor system for microgrids in Africa. In *2021 IEEE AFRICON* (pp. 1–5). IEEE. https://doi.org/10.1109/AFRICON51333.2021.9570970

Mwammenywa, I. (2024). A Novel Autonomous and Real-time Load Monitoring and Control System in Microgrids Based on Fuzzy Logic Control and LoRa Wireless Communication (Doctoral dissertation, Universitätsbibliothek). https://doi.org/10.17619/UNIPB/1-2019

Mwammenywa, I., Petrov, D., Holle, P., & Hilleringmann, U. (2022). LoRa transceiver for load monitoring and control system in microgrids. In *2022 International Conference on Engineering and Emerging Technologies (ICEET)* (pp. 1–5). IEEE. https://doi.org/10.1109/ICEET56468.2022.10007274

PwC. (2016). *Electricity Beyond the Grid: Accelerating Access to Sustainable Power for All.* PwC: Global Power & Utilities. Retrieved from: https://www.esi-africa.com/wp-content/uploads/2016/05/electricity-beyond-grid.pdf

Serban, I., Cespedes, S., Marinescu, C., Azurdia-Meza, C. A., Gómez, J. S., & Hueichapan, D. S. (2020). Communication requirements in microgrids: A practical survey. *IEEE Access, 8*, 47694–47712.

Tightiz, L., & Yang, H. (2020). A comprehensive review on IoT protocols' features in smart grid communication. *Energies, 13*(11), 2762.

Wang, W., Xu, Y., & Khanna, M. (2011). A survey on the communication architectures in smart grid. *Computer Networks, 55*(15), 3604–3629.

4. Societal Perspectives

4.1 Energy Development in East Africa

Tobias Klaus

4.1.1 Historical Context and Post-Colonial Energy Paradigms

After decolonization in the 1960s, East African countries were left with energy systems primarily designed to serve colonial administrative and economic interests. Infrastructure and access for the broader population (Archibong & Afolabi, 2023) were limited. Yet, the region is endowed with abundant energy resources, including fossil fuels like oil, coal, and gas, as well as hydropower, geothermal, solar, and wind. For Southern Africa, it has been noted that the development of these resources has been fragmented and shaped by both domestic and international political and economic forces (Power et al., 2016).

The energy landscape in the states of the East African Community (EAC) (Kenya, Uganda, Tanzania, South Sudan, Rwanda, Burundi, and the DRC) has over the years evolved from fragmented colonial-era infrastructure systems towards more integrated regional systems. Still, significant disparities persist, which are reflected in very unequal access to those infrastructures. In the electricity sector, initial post-independence efforts focused on large-scale hydropower and fossil fuels, with rural electrification progressing only on a small scale until the 2000s. Biomass remained dominant—meeting 80 % of total energy needs in 2014. The vast majority (97 %) of the population still rely on traditional sources like wood and charcoal (SID, 2018).

4.1.2 Evolution of Energy Supply

More recently, in the second decade of the 21st century, the member states of the EAC roughly doubled their total electricity generation capacity, reaching about 10.4 GW by 2024, with Kenya and the DRC accounting for 7 GW of the overall amount.

In comparison, the overall generation capacity at independence amounted to only about 850 MW, with the population having grown more than five-fold since then. More than 80 % of this expansion was driven by the deployment of renewable energy conversion facilities, in particular hydropower, geothermal, wind, and solar (cf. RES4Africa Foundation, 2023; Corvaglia et al., 2023). Access to electricity ranges from 5.4 % in South Sudan to 76.2 % in Kenya (Tab. 4).

Tab. 4: Access and generation of electricity in East Africa (author's own compilation)

		Population			**Electricity Generation Capacity**			**Access to Electricity**
	Year of independence	at independence (World Bank)	2024 (World Bank)	growth in %	at independence (estd.)	2023 in MW (Statista, SACREEE)	growth in %	2023 in % (World Bank)
Burundi	1962	2,907,943	14,047,786	483	5	126.65	2.533	11.6
DRC	1960	15,264,846	109,276,265	716	570	2819	495	22.1
Kenya	1963	8,627,971	56,432,944	654	102	3417.74	3.351	76.2
Rwanda	1962	3,186,025	14,256,567	447	5	300.45	6.009	63.9
Somalia	1960	2,876,331	19,009,151	661	5	155.46	3.109	50.3
South Sudan	2011	10,288,109	11,943,408	116	22	212.37	965	5.4
Tanzania	1961	10,187,886	68,560,157	673	21	2008.74	9.565	48.3
Uganda	1962	8,092,231	50,015,092	618	120	1324.78	1.104	51.5
Sum		**61,431,342**	**343,541,370**	**559**	**850**	**10365.19**	1.219	

Hydropower still accounts for the majority of the region's electricity mix, but diversification into other renewables has advanced significantly, particularly in Kenya, which has established extensive regulatory frameworks and ambitious targets for the integration of renewable energy (Corvaglia et al., 2023). While only in Somalia, South Sudan, and Tanzania the percentage of renewables is far below 50 %, the share in Rwanda is about even, and in the remaining countries it ranges between 65 % (Burundi) and 100 % (DRC) (Statista, SACREEE).

Despite a significant deployment of renewable generation facilities, fossil fuels remain significant, particularly in Tanzania and Uganda, where oil and gas exploration and export projects like the East African Crude Oil Pipeline (EACOP) are underway (GEM Wiki, 2025; Daily News, 2025). These projects are politically and economically significant, yet they have raised concerns regarding governance, social displacement, and the effects on the environment (GEM Wiki, 2025; Daily News, 2025).

4.1.3 Development of Energy Demand

Population growth, urbanization, and economic development have driven a steady increase in energy demand across East Africa. However, access to electricity remains extremely low—below 50 % in many countries—with rural areas being particularly underserved (GEM Wiki, 2025). The growing demand is further intensified by industrialization efforts and the expansion of services and infrastructure (Energy Charter Secretariat, n.d.).

Given the technical, social, and environmental challenges and the resulting costs of extending national grids, off-grid solar and microgrid solutions have become an important instrument for expanding access, especially in remote and rural areas. Uganda, for example, is leveraging its renewable potential to promote off-grid solutions, which are seen as both a means to increase access and a tool for peace-building in conflict-affected areas (GEM Wiki, 2025), exemplified by the GIZ Pro Mini Grids project (GIZ, 2022).

4.1.4 Political and Social Institutions: Drivers and Barriers

Energy sector development in East Africa is dependent on factors rooted in the political sphere. Government priorities, regulatory environments, and the influence of international investors have shaped the pace and nature of energy transition (Archibong & Afolabi, 2023; Danish Institute for International Studies, 2023; Power et al., 2016). The Danish Institute for International Studies concluded that Tanzania's energy policy has shifted in response to broader development strategies. Renewables were often neglected due to lack of political will or investment incentives (Danish Institute for International Studies, 2023).

Kenya's relatively sophisticated regulatory framework—including feed-in tariffs, grid codes, and liberalization measures—has attracted private investment and accelerated renewable deployment, whereas the absence of regional regulatory planning has hindered coordinated progress across the region (Corvaglia et al., 2023).

Access to modern energy is considered essential for socio-economic development, poverty reduction, and improved livelihoods (Energy Charter Secretariat, n.d.). However, large-scale projects have sometimes led to displacement, environmental degradation, and social tensions, especially when local communities are not adequately consulted or compensated (GEM Wiki, 2025; Hategekimana, 2025). Even for small projects, acceptance and sustainability are influenced by local communities and traditional institutions, whose support is vital for project implementation (Mwangi, 2016).

For a structured analysis of resource governance, a good understanding of the links between the relevant institutions, actors, and drivers is of crucial importance. Here, the *Institutional Analysis and Development Framework* by Ostrom et al. (see, e.g., Ostrom 2009) and related variants of this framework have proven useful. They provide a structured way to examine how individuals interact within various governance systems, such as managing natural resources or creating public policy, and they have provided an analytical basis for several of the sections in this book.

In those frameworks, the empirical data is structured according to the *action arenas*, the relevant *actors*, *rules*, and *outcomes*. For our present purposes, and on a general level, data points can thus be assigned within Tab. 5 below, and social, political, and environmental questions can be treated as resulting from this matrix.

Tab. 5: The IAD framework of microgrids

Action Arena	Actors	Rules-in-Use	Outcomes
Policy Formulation	EAC governments, international organizations (e.g., IRENA, African Development Bank, World Bank), donors	Liberalization policies; renewable targets (Africa Clean Energy Corridor) (IRENA, 2015)	Regional master plans (e.g., EAPP); \$2.7B/year generation investment (Falchetta et al., 2020)
Project Implementation	Utilities, communities, NGOs	Community engagement norms; public-private partnerships	Higher project sustainability where local participation prioritized (IRENA, 2015; SID, 2018)
Resource Governance	Regional bodies (EAC), national agencies	Cross-border treaties; harmonized regulations (e.g., PIDA energy projects) (IRENA, 2015)	Shared infrastructures (e.g., Ethiopia-Sudan 500 kV line); reduced costs (IRENA, 2015)

4.1.5 Social and Political Problems

Energy Inequality. Despite progress, a significant portion of the population remains without equitable and sustainable access to energy systems, which in consequence exacerbates social and economic inequalities (GEM Wiki, 2025), including gender inequalities.

Governance and Corruption. The management of energy resources, particularly oil and gas, as well as large hydropower, has been linked to patronage systems and corruption, undermining equitable development and peace-building, especially in fragile states like South Sudan (cf. e.g., GEM Wiki, 2025).

Environmental and Social Risks. Large infrastructure projects have triggered environmental concerns and social displacement, often without adequate mitigation or compensation mechanisms (GEM Wiki, 2025; Hategekimana, 2025), affecting the rural poor in particular.

Regulatory and Institutional Fragmentation. The lack of harmonized regional planning and regulatory frameworks obstructs the efficient deployment of renewables and the integration of the energy market (Corvaglia et al., 2023).

4.1.6 Conclusion

Since decolonization, the evolution of energy supply and demand in the EAC is deeply interwoven with political and social institutions, which will be further explained in the following sections. While significant challenges persist—especially regarding access, governance, and sustainability—the region's growing emphasis on renewables, regional cooperation, and community engagement offers promising avenues for inclusive and sustainable energy development. The analyses in this chapter highlight the complex, adaptive interplay between formal and informal institutions, suggesting that continued progress depends on strengthening institutional capacity, fostering synergies, and addressing persistent inequalities.

4.1.7 References

Archibong, E. I., & Afolabi, A. P. (2023). From colonial exploitation to renewable transition: a critical analysis of Africa's energy paradigm. *European Journal of Sustainable Development Research*, *7*(4), em0236.

Corvaglia, F., Formicola, G., Grisi, T., Guzzo, D., Ibtissem, H., Ledauphin, M., & Serao, M. (2023). *A Challenging Transition: Exploring East Africa's Last 10 Years of Renewable Energy*. RES4Africa Foundation.

Daily News. (2025). Unlocking East Africa's energy wealth, sustainability. *Daily News*. https://dailynews.co.tz/unlocking-east-africas-energy-wealth-sustainability/

Danish Institute for International Studies. (2023). *The Political Economy of Energy Production in Post-independence Tanzania*. DIIS Working Paper 2023, No. 1.

Energy Charter Secretariat. (n.d.). *Energy in the East African Community*. Brussels: Energy Charter Secretariat.

Falchetta, G., Hafner, M., & Tagliapietra, S. (2020). Pathways to 100 % electrification in East Africa by 2030. *The Energy Journal*, *41*(3).

GEM Wiki. (2025). Power sector transition in East Africa (EACOP region). https://www.gem.wiki/Power_Sector_Transition_in_East_Africa_(EACOP_region)

GIZ. (2022). *Promotion of Mini Grids for Rural Electrification (Pro Mini Grids)*. GIZ. https://www.giz.de/de/downloads/giz2022-en-mini-solar-grids.pdf

Hategekimana, S. (2025). Conceptualising energy geographies in East Africa. *Energy for Sustainable Development*, *72*, 83–87.

International Renewable Energy Agency (IRENA). (2015). *Africa Clean Energy Corridor: Analysis of Infrastructure for Renewable Power*. IRENA.

Mwangi, E. (2016). Community participation in renewable energy projects: a case study of Kenya. *Renewable Energy, 92*, 307–315.

Ostrom, E. (2009). *Understanding Institutional Diversity*. Princeton: Princeton University Press.

Power, M., Newell, P., Baker, L., Bulkeley, H., Kirshner, J., & Smith, A. (2016). The political economy of energy transitions in Mozambique and South Africa: the role of the rising powers. *Energy Research and Social Science, 17*, 10–19.

RES4Africa Foundation. (2023). *Charting a Sustainable Energy Course for East Africa*. https://res4africa.org/news/2023/in-between-challenges-and-potential-charting-a-sustainable-energy-course-for-east-africa/

SACREEE. (n.d.). The Democratic Republic of Congo. https://www.sacreee.org/index.php/member- state/democratic-republic-congo

Society for International Development (SID). (2018). *Energy for Whom? Scenarios for Eastern Africa*. Berlin: Heinrich-Böll-Stiftung.

Statista. (n.d.). https://www.statista.com/

World Bank. (n.d.). World Bank Open Data. https://data.worldbank.org/

4.2 Socio-Cultural and Institutional Economic Analyses of Microgrids

4.2.1 Introduction

Burkhard Hehenkamp

Microgrids stand at the crossroads of technology and society, offering a promising pathway to sustainable rural electrification across East Africa (Ahlborg & Hammar, 2014; Motjoadi et al., 2020). However, their integration into rural communities is not merely a technical challenge but a socio-economic and cultural endeavor (de-Bruijn & Herder, 2009; Righi & Saurin, 2015). This section explores how socio-cultural dynamics and institutional arrangements influence microgrid governance, with a particular focus on how communities govern decentralized energy systems in practice—not just on paper.

Regulatory frameworks are the scaffolding with which microgrid operations are built, shaping everything from tariff structures to operational protocols (Namujju et al., 2023). These policies not only determine the viability of microgrid projects but also influence how they are perceived and utilized within rural communities. A deeper understanding of these frameworks reveals how policy can be tailored to foster both institutional stability and community acceptance. In this vein, Section 4.2.2 examines to what extent institutional factors influence how much households connected to a microgrid in Tanzania are willing to pay for electricity.

Sustainability, however, cannot be fully explained by institutional and policy structures alone. Microgrids are best understood as complex socio-technical systems (STS)—ensembles of technology, human agency, institutions, and cultural practices (Anderies et al., 2004). Integrating STS theory with institutional economics and socio-cultural analysis allows for a holistic examination of how technical capacity, informal institutions, and perceptions of ownership interact to influence governance outcomes. This interplay is explored in more detail in Section 4.2.3.

Equally important are the ownership structures that underpin microgrid operations. Whether community-managed, privately operated, or governed through public-private partnerships, ownership influences everything from cost recovery to social legitimacy. Ownership determines who holds decision-making power, how benefits are distributed, and how responsive the system is to local needs (Ngoti, 2024b). Section 4.2.4 delves into these relationships.

Lastly, regulatory and policy frameworks are limited in their efficacy if they are not adequately enforced. In such cases, informal institutions take their place. This is where Section 4.2.5 digs in. It examines the structure of governance networks through a social network analysis (SNA). By mapping perceived authority relationships and identifying structural power asymmetries, the SNA approach uncovers how governance actually functions on the ground (Carlsson & Sandström, 2008; Chiesi, 2006). Our findings highlight the role of often unseen informal infrastructures that sustain off-grid systems, offering important insights into how decentralized energy governance might be more effectively supported, integrated, and reformed.

4.2.2 Do Institutional Factors Affect Households' Willingness to Pay for Microgrid Electricity in Tanzania?

Henrietta Acquah-Swanzy

4.2.2.1 Introduction

The sustainability of microgrids has emerged as a critical focus for policymakers, development practitioners, and scholars concerned with rural electrification and energy access in SSA (Ahlborg & Hammar, 2014). As decentralized energy systems are increasingly promoted as viable pathways for achieving universal electrification (IEA, 2022), attention has turned to the economic viability of these systems, and in particular their ability to recover costs and ensure long-term service provision (Tenenbaum et al., 2014). Central to this is the household's *willingness to pay* (WTP) for electricity from microgrids, directly influencing revenue sufficiency and the potential for scaling and sustaining microgrid operations. In this study, we examine factors influencing the WTP among microgrid users, focusing on institutional factors and perceptions.

The WTP reflects the economic capacity of households and their valuation of electricity services relative to other expenditures (Abdullah & Jeanty, 2011). Numerous studies have identified socio-economic determinants of the WTP, such as income (Boait et al., 2019), household characteristics (Bukari et al., 2021), education, and gender (Ondraczek, 2013). While these factors have been widely examined, there remains limited empirical (especially quantitative) work on how institutional features specific to microgrid design and operation influence the WTP (Trotter et al., 2017). However, such institutional variables can affect user experiences, perceptions of fairness, and transaction costs, impacting payment behaviour and WTP (Trotter et al., 2017).

This research gap is particularly salient for SSA, where innovations such as prepaid metering, mobile money integration, and flexible tariff structures are increasingly embedded in decentralized energy systems (Muchunku et al., 2018). Studies of grid electrification have shown that billing design and payment methods can directly affect payment compliance, affordability perceptions, and customer satisfaction (Jack & Smith, 2020). However, there is limited empirical evidence on how these institutional arrangements influence WTP in the context of microgrids and other off-grid solutions. Understanding these dynamics is crucial for designing financially sustainable, user-centred, decentralized electricity systems. This study aims to address this gap and provide actionable insights for energy access policy and microgrid operations by examining the role of billing method and payment modality alongside socio-economic determinants.

In this study, we investigate households' WTP for microgrid electricity in Tanzania, addressing institutional factors embedded in the design and operation of microgrid systems. We draw on household survey data and employ a logistic regression to estimate the effects of institutional and socio-economic predictors on the WTP. By integrating institutional analysis with socio-economic modelling, this chapter contributes to a more nuanced understanding of demand-side dynamics in decentralized energy transitions.

4.2.2.2 Study Framework

4.2.2.2.1 Willingness to Pay for Microgrid Electricity

The WTP is the maximum amount a household is willing to pay for access to and continued electricity use provided by a microgrid system. This reflects the economic capacity and perceived value of service. This concept is closely linked to the ability to pay, which is directly linked to the income of users. In rural Tanzania, low disposable income limits the affordability of electricity tariffs. Boait et al. (2019) studied a solar-powered microgrid on a Lake Victoria island and found that the operator's tariff of 3500 Tanzanian Shillings per kWh was unaffordable, given the average daily income of $1.90 per person. Despite this, households expressed a strong WTP within their means, driven by the perceived benefits of electricity for healthcare and economic activities (Boait et al., 2019).

4.2.2.2.2 Institutional Factors Influencing WTP

Institutional factors reflect rules, structures, systems, and service delivery mechanisms related to how the microgrid operates or is governed. They shape the operational environment and user interface of the microgrid system, influencing trust, transaction costs, perceived fairness, and user satisfaction, all of which mediate the WTP. For instance, a study based on eight months of fieldwork conducted in six rural communities in Tanzania reveals how private-sector-led microgrid electrification, while expanding access, may reproduce energy injustices through cost allocation and exclusionary practices (Mottram, 2022). Institutional features are not neutral technical choices; they embed normative and distributive outcomes that determine who benefits from electrification and under what conditions. Tariff design, billing methods (e.g., flat vs. consumption-based), and payment modalities (e.g., mobile money vs. manual payment) directly influence transaction costs, perceived legitimacy, and the practical feasibility of maintaining a connection. Fig. 43 shows some institutional and socio-economic factors and the mediating factors on the WTP. Institutional factors may have a direct effect on the WTP but can also indirectly affect the WTP through mediating perceptions of microgrid users.

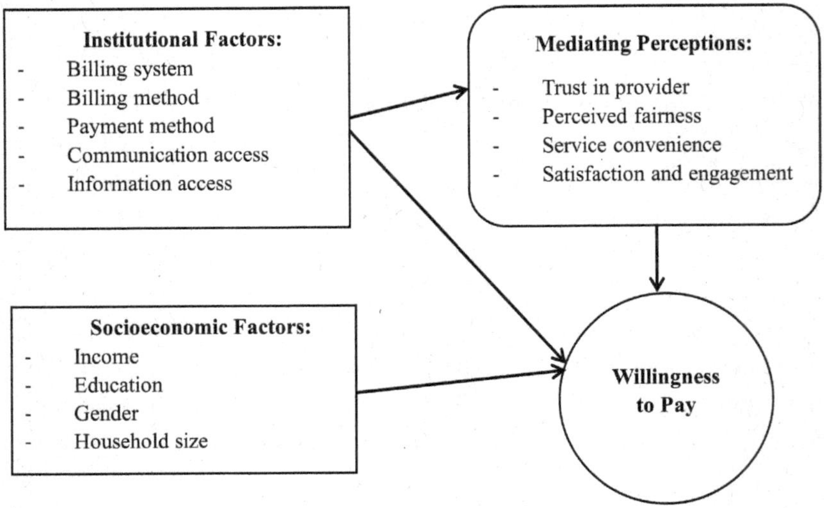

Fig. 43: Factors influencing willingness to pay showing the mediating perceptions.

4.2.2.3 Study Method

4.2.2.3.1 Survey and Household Characteristics

This study utilizes data from the 2023 household interdisciplinary survey conducted under the A:RT-D Grids project, which covered 25 rural communities in Tanzania. These communities are served by 11 operational microgrids that vary in ownership models, technological configurations, and management structures. The survey collected detailed demographic and socio-economic information at the household level, along with data on electricity access and utilization patterns, particularly in relation to household interactions with microgrid institutions.

A total of 1,392 households were surveyed, of which 904 (approximately 68 %) reported having electricity access in their homes. For these connected households, additional data were collected on institutional features of electricity provision. These included the type of billing system (prepaid or postpaid), billing method (flat rate or consumption-based), preferred payment modality (mobile money or cash), and user-provider communication channels.

The survey also incorporated measures of community perceptions regarding the fairness of service provision, responsiveness of electricity providers, affordability of electricity, and procedural transparency in obtaining connections. These variables allow for a multidimensional analysis of institutional trust and service delivery satisfaction.

4.2.2.3.2 *Measurement of Key Variables*

The primary outcome variable, the willingness to pay (WTP) for electricity services, was elicited through the question: »How much are you willing to pay as your monthly electrical bill (in local currency)?« Responses were categorized into five ordinal bins: »not_more_than_5000,« »5001_to_10000,« »10001_to_20000,« »20001_to_50000,« and »50001_to_80000« (TZS). Distributional analysis indicated substantial right-skew, with over 93 % of respondents concentrated in the two lowest categories. The WTP was recoded into a binary variable to address this imbalance and support robust modelling: 0 for a WTP $\leq$ 10,000 and 1 for a WTP > 10,000. This threshold reflects a meaningful distinction between minimal and substantive payment commitment in the context of the study.

A similar binning structure captured a related indicator, the ability to pay (ATP), and how much households currently pay for electricity in a month (CuPayment). Cross-tabulation and chi-square tests confirmed a strong positive association between the WTP and the ATP as well as between the WTP and CuPayment, suggesting that stated willingness closely tracks perceived financial capacity.

All variables were harmonized and cleaned prior to analysis. Non-informative responses such as »I don't know« and placeholder codes (e.g., 999) were coded as missing. Tab. 6 presents summary statistics for all variables included in the analysis, limited to households with an electricity connection (N = 904).

4.2.2.3.3 *Econometric Model*

Recoding of the WTP into a binary variable allowed for analysis using binary logistic regression focusing on the critical threshold of 10,000 TZS. The regression model is specified as

$$\mathit{logit}\ (Pr\ (WTP_i > 10000)\,) = \beta_0 + \sum_k \beta_k X_{\{ik\}} + \sum_m \gamma_m Z_{\{im\}},$$

where $X_{\{ik\}}$ represents socio-economic and institutional covariates, and $Z_{\{im\}}$ denotes perception mediators. We estimate three models: a full model including socio-economic, institutional, and mediating variables; a model including institutional factors and mediating factors; and a last model that tests socio-economic factors with institutional factors. Tab. 7 presents all marginal effects of the binary logistic regression for all three models.

4.2.2.4 Empirical Results and Discussion

4.2.2.4.1 *Effect of Billing System on the WTP*

The most significant predictor across all models is using a prepaid billing system, which consistently increases the probability of WTP > 10,000. The highest effect is seen in Sub-Model 1 in Tab. 7, where we estimate the WTP with institutional variables and the mediating factors. Prepaid systems likely enhance the WTP by giving households greater control over electricity usage and expenditure, reducing the perceived financial risk associated with postpaid billing.

4.2.2.4.2 *Effect of Multiple Information Access Sources on the WTP*

Information access count exhibits a consistent negative effect on the WTP across all models, reducing the probability by 5.3 % per additional access point in the Full Model, 6.0 % in Sub-Model 1, and 3.0 % in Sub-Model 2. This unexpected result may reflect information overload, where excessive engagement leads to confusion or scepticism about the microgrid's benefits. Alternatively, frequent information-seeking might indicate underlying distrust or dissatisfaction, which could suppress the WTP. Similar patterns have been observed in rural electrification projects, where poorly designed information campaigns can undermine user confidence (Bhattacharyya, 2013).

4.2.2.4.3 *Effect of Perceived Fairness on the WTP*

Our results show that perceived microgrid fairness, a mediating factor, significantly and positively influences the probability of a higher WTP. These effects are seen in the Full Model (AME = 0.064, $p < 0.05$, OR = 4.36) and Sub-Model 1 (AME = 0.069, $p < 0.01$, OR = 5.04). Households perceiving the microgrid as fair are more likely to pay > 10,000, underscoring the role of trust in encouraging financial commitment.

4.2.2.4.4 *Effect of Socio-economic Factors on the WTP*

Socio-economic factors have a limited impact. Gender (male) shows a marginally negative effect in the Full Model, which is insignificant in Sub-Model 2. Income source (business/trade) is significant only in Sub-Model 2, indicating a modest positive effect for business households. Education, age, and household size are consistently insignificant, suggesting that demographic factors are less critical than institutional and behavioural drivers in this context. This aligns with prior studies on microgrid adoption, which often find that structural factors outweigh socio-economic characteristics in shaping payment behaviour (Alstone et al., 2015).

4.2.2.4.5 *Institutional Design and Trust in Microgrid Systems*

The significant positive effect of prepaid billing systems on the WTP highlights the importance of institutional design in microgrid deployment. By allowing users to pay upfront and monitor usage, prepaid systems align with the financial realities of low-income households in developing regions, where income is often irregular and unpredictable. This finding is consistent with evidence from rural electrification initiatives in SSA, where prepaid metering has been shown to reduce default rates and increase user satisfaction by providing transparency and control (Jack & Smith, 2020). The high probability effect in Sub-Model 1 further emphasizes the transformative potential of this institutional feature, suggesting that microgrid operators should prioritize prepaid systems to encourage higher financial commitments.

Equally important is the role of perceived fairness, which significantly boosts the WTP in models that include mediating factors. Households perceiving the microgrid as fair are 6.4 % to 6.9 % more likely to pay > 10,000 ($p < 0.05$), reflecting the critical role of trust in community-based energy systems. This finding aligns with research on renewable energy adoption, where trust in service providers and perceived equity in resource distribution are key determinants of user engagement (Walker et al., 2010). In microgrid contexts, trust can be fostered through transparent pricing, participatory decision-making, and equitable service delivery, all enhancing users' willingness to invest financially (Sovacool & Dworkin, 2015). The absence of this variable in Sub-Model 2 likely contributes to its lower explanatory power, underscoring the need to integrate trust-related factors into microgrid policy design.

The negative effect of digital payments in Sub-Model 2 contrasts with the positive impact of prepaid systems, highlighting a tension in institutional design. While digital payments are often promoted for their efficiency, their deterrent effect on the WTP may reflect challenges such as limited digital literacy, high transaction costs, or lack of trust in digital platforms—issues well-documented in rural energy markets (Alstone et al., 2015). This suggests that institutional innovations must be carefully tailored to local contexts, ensuring that payment systems are accessible and trusted by users.

4.2.2.4 Conclusion

The results underscore the importance of institutional and mediating factors, such as prepaid billing systems and perceived fairness, in driving the WTP for microgrid services. In contrast, socio-economic factors play a limited role. The model involving institutional and mediating factors offers the best predictive performance, making it a valuable tool for targeting interventions, though the full model, which includes socio-economic factors, provides a more holistic understanding. These findings highlight actionable strategies for microgrid policymakers, including promoting prepaid billing, enhancing fairness perceptions, and addressing barriers to digital payment adoption. Future research should focus on improving predictive accuracy for rare events and exploring the behavioural mechanisms underlying these effects.

Tab. 6a: Descriptive statistics: Dependent Factor. *Notes (Tab. 6a-d): Summary from available sample data of those connected to microgrid. Numbers in brackets are the standard deviations. Service rating score uses PCA to combine staff expertise and query response ratings. WTP and ATP are not independent (Prob > |z| = 0.0000), like WTP and current monthly payment. Information Access Mean 2.70 (1.70).*

Characteristic		Frequency	Percentage
WTP	Low	819	90.60
	High	85	9.40
ATP	Low	811	89.71
	High	93	10.29
Currently Paying	Low	794	88.03
	High	108	11.97

Tab. 6b: Descriptive statistics: Socio-economic Factors.

Characteristic		Frequency	Percentage
Gender	Female	309	34.14
	Male	596	65.86
Average Age	41.63 (12.34)		
Average Household Size	5,40 (2.70)		
Education	None	27	2.98
	Primary	518	57.24
	Secondary >	360	39.78
Income Source	Agriculture	723	79.98
	Business/Trading	117	12.94

Characteristic		Frequency	Percentage
	Formal salary	56	6.19
	Other	8	0.88

Tab. 6c: Descriptive statistics: Institutional Factors.

Characteristic		Frequency	Percentage
Billing System	Postpaid	546	60.67
	Prepaid	354	39.33
Billing Method	Consumption Based	628	69.55
	Flat rate	275	30.45
Payment Type	Cash	500	55.56
	Digital	308	34.22
	Both	92	10.22

Tab. 6d: Descriptive statistics: Mediating Factors.

Characteristic		Frequency	Percentage
Perceived Fairness	No	25	2.87
	Yes	847	97.13
Payment Mood Comfort	No	106	11.76
	Yes	795	88.24
Average Service Rating	-0,06 (1.40)		

Tab. 7: Binary Logistic Average Marginal Effects (AME) for WTP > 10,000 *Notes: Marginal effects (AME) represent the change in probability of WTP > 10,000 for a one-unit change in the predictor (continuous) or a discrete change from the base level (categorical), with standard errors in parentheses. Significance levels: *** p < 0.01, ** p < 0.05, * p < 0.10. Full model: N = 828, Sub-Model 1 (institutional and mediating factors): N = 828, Sub-Model 2 (institutional and socio-economic factors): N = 897.*

Predictor	Full Model AME (SE)	Sub-Model 1 AME (SE) (Inst. + Med.)	Sub-Model 2 AME (SE) (Inst. + Socio.)
Socio-economic Factors			
Gender (Base: Female)			
Male	-0.031 (0.019) *	–	-0.023 (0.019)
Age	0.001 (0.001)	–	0.001 (0.001)
Household Size	0.000 (0.003)	–	0.003 (0.003)
Education (Base: None)			
Primary	0.009 (0.035)	–	0.012 (0.040)
Secondary or Higher	0.047 (0.039)	–	0.030 (0.042)
Income Source (Base: Agriculture)			
Business/Trade	0.036 (0.026)	–	0.054 (0.027) *
Formal Salary	0.018 (0.038)	–	0.064 (0.047)
Other	-0.024 (0.045)	–	0.008 (0.063)
Institutional Factors			
Billing Method (Base: Other)			
Flat Rate	-0.013 (0.027)	-0.030 (0.024)	-0.007 (0.028)
Billing System (Base: Postpaid)			
Prepaid	0.313 (0.059) ***	0.368 (0.053) ***	0.322 (0.051) ***
Payment Type (Base: Cash)			
Digital	-0.013 (0.040)	-0.042 (0.039)	-0.110 (0.043) **

Predictor	Full Model AME (SE)	Sub-Model 1 AME (SE) (Inst. + Med.)	Sub-Model 2 AME (SE) (Inst. + Socio.)
Both	-0.025 (0.072)	-0.036 (0.078)	-0.062 (0.083)
Information Access Count	-0.053 (0.009) ***	-0.060 (0.009) ***	-0.030 (0.008) ***
Mediating Factors			
Service Rating	-0.009 (0.007)	-0.007 (0.007)	–
Mode of Payment Comfort (Base: No)			
Yes	-0.023 (0.026)	-0.021 (0.025)	–
Perceived Microgrid Fairness (Base: No)			
Yes	0.064 (0.021) ***	0.069 (0.019) ***	–

4.2.3 Technical Capacity and Hierarchical Dynamics in Community-led Microgrids

Lillian Donna Namujju, Burkhard Hehenkamp, Teddy Mangeni

4.2.3.1 Introduction

Community microgrids serve as decentralized energy solutions for off-grid populations, yet their sustainability remains a persistent challenge. As *common-pool resources* (CPRs), microgrids exhibit characteristics of *rivalry* (where excessive demand can strain system capacity) and difficulty in *exclusion* (as weak enforcement mechanisms allow unauthorized or unpaid usage) (Ostrom, 1990). Similar to other CPRs, such as community water systems or fisheries, microgrids require robust governance structures to prevent overuse, free-riding, and financial instability. Without clear rules, enforcement mechanisms, and accountability structures, community-led microgrids often experience system failures, poor financial performance, and eventual collapse (Akinyele et al., 2018).

Beyond governance, microgrids are also complex *socio-technical systems* (STS), where technological performance is deeply intertwined with human behaviour, institutional arrangements, and knowledge distribution (Namujju, 2024). In community microgrids, sustainability depends not only on technical skills for maintenance and repairs but also on social structures that define who controls these decisions and how responsibilities are distributed. A key challenge is that technical knowledge is often concentrated in a few individuals while users lack both technical expertise and awareness of their oversight role (Hatziargyriou et al., 2016). This power asymmetry limits accountability, weakens financial discipline, and leads to inefficient system management, ultimately undermining the long-term viability of the microgrid (Boche et al., 2022; Ottenburger et al., 2024).

The dominant discourse on microgrid sustainability focuses on technical capacity-building as a solution—assuming that equipping operators with technical skills will improve performance (Bogere et al., 2023; Nwanevu et al., 2024; Sulaeman et al., 2021). However, this overlooks the governance hierarchies and institutional dynamics that shape how technical decisions are made and enforced. In many cases, the host community lacks the authority to challenge technicians' decisions, and community members, unaware of their rights, fail to demand transparency and accountability in system management. This research shifts the focus beyond technical capacity deficits to the power relationships between technicians, local government representatives, community committees, and users, examining how these interactions influence financial management, maintenance practices, and overall sustainability.

This study examines a *community microgrid* in Tanzania as a case study to explore these governance and technical challenges. By analysing the interactions between technicians, the supervisory District Energy Office, the Village Energy Committee, and the broader community, the study investigates how power asymmetries, lack of stakeholder engagement, and technical knowledge gaps contribute to microgrid sustainability failures. We specifically seek to address the following key questions: (1) How do hierarchical governance structures interact with technical capacity to shape microgrid operations and long-term sustainability? (2) How can technical training be redesigned to both enhance skills and democratize decision-making power in microgrid governance?

This study expands the microgrid sustainability discourse beyond technical skill gaps, integrating CPR theory to explain how governance failures contribute to financial and operational instability. It also applies STS theory to explore how power asymmetries in technical knowledge distribution shape decision-making and oversight capacity in community-led microgrids. By integrating insights from governance theory and socio-technical systems analysis, this study provides a multi-dimensional approach to improving the sustainability of community microgrids.

4.2.3.2 Institutional and Technical Dimensions of Microgrid Governance

This study is grounded in CPR theory and STS theory, two complementary frameworks that provide insight into the governance and sustainability challenges of community microgrids. CPR theory explains the institutional and governance failures that arise when shared resources are poorly regulated (Ostrom, 2002; Roy et al., 1990), while STS theory highlights the interplay between technology, human behaviour, and institutional structures in shaping microgrid performance (Anderies et al., 2004; Cooper and Foster, 1971). Together, these theories provide a robust foundation for analysing how hierarchical governance structures and technical capacity interact to influence microgrid operations and long-term sustainability.

4.2.3.2.1 CPR Theory and Hierarchical Governance Structures in Microgrids

CPR theory, when applied to microgrid governance, highlights the challenges of managing shared resources, particularly in terms of governance failures, weak enforcement mechanisms, and resource mismanagement (Gollwitzer, 2014; Nathan et al., 2022). However, these challenges are not solely institutional; they are worsened by hierarchical governance structures and disparities in technical capacity.

A key governance issue in community microgrids as CPRs is power asymmetry in decision-making, where technical and operational control is concentrated among a few actors, undermining collective governance and reinforcing CPR management failures (Cox et al., 2009). When external or local agencies dominate decision-making, it disrupts collective governance, which CPR theory identifies as essential for preventing resource mismanagement (Ostrom, 1990). Decision-making authority remains centralized, excluding community stakeholders from participating in critical governance functions. Consequently, the community's role in rule enforcement is diminished, limiting its ability to prevent free-riding, regulate consumption, or ensure financial sustainability (ibid). This weakens compliance mechanisms, making it difficult for community members to challenge inefficiencies, monitor technical operations, or hold decision-makers accountable (Lozny, 2023).

4.2.3.2.2 *Technological Complexity, Technical Capacity Disparities in Governance*

Technical capacity disparities in microgrid governance concentrate expertise within a few actors, reinforcing hierarchical and technocratic structures that marginalize community stakeholders (Kavanagh, 2019; Pereira and Quintana, 2002). As microgrids operate within complex socio-technical systems, those with technical knowledge gain disproportionate decision-making power, often going unchallenged even when inefficiencies or failures arise (Cegarra-Navarro et al., 2014). This power asymmetry is further entrenched by exclusive access to system data and diagnostics, limiting transparency and sidelining less technically proficient groups, such as community committees, from meaningful oversight (Lightfoot and Wisniewski, 2014). As a result, governance shifts from participatory to technocrat-led, weakening local ownership and accountability.

Further, technical capacity gaps contribute to resource management failures in several ways. Communities and local governance bodies with limited technical expertise often struggle with monitoring system performance, enforcing financial discipline, and ensuring effective oversight of operational decisions. This lack of expertise fosters over-reliance on external agents for critical decision-making, further reducing local autonomy and weakening community ownership of the microgrid (Slough et al., 2021). As a result, financial mismanagement, technical inefficiencies, and an inability to enforce sustainable consumption practices become recurring challenges, ultimately threatening the long-term sustainability of community-led microgrids.

4.2.3.3 Data, Tools, and Approach

This study was conducted in a Tanzanian community microgrid. It employs a qualitative case study approach, using semi-structured interviews to explore governance challenges, technical capacity disparities, and hierarchical decision-making structures in the community. Data was collected from 15 key respondents, including 2 Village Electricity Committee (VEC) members, 2 local government representatives, the microgrid technician, and 10 microgrid consumers.

Interviews focused on respondents' perceptions of microgrid governance, their level of technical knowledge, interactions with technicians, and their role in decision-making and oversight. The data was transcribed and systematically coded using MaxQDA, allowing for thematic analysis of governance failures, power asymmetries, and barriers to technical participation. This methodology provides in-depth insights into the socio-technical and institutional dynamics affecting microgrid sustainability.

4.2.3.4 The Case of the Tanzanian Community Microgrid

The microgrid in rural Tanzania was established in 2015 as a public-private partnership (PPP) between the Tanzanian government and an independent Australian energy developer. The decision to implement the energy project was initiated by the Ministry of Energy due to strong advocacy from the local Member of Parliament. In 2019, the system transitioned to community ownership, with the Village Electricity Committee (VEC) assuming responsibility for management, tariff-setting, and revenue collection. However, this transition was informal, lacking official handover documentation, clear governance structures, and strong enforcement of payment rules. Despite its community-managed status, the oversight role remained with the District Energy Office, further complicating local governance. Since the transition, the microgrid has struggled with a number of issues highlighted in Tab. 8, as derived from the field data.

Tab. 8a: Case Study Microgrid—Survey themes and prevailing stakeholder sentiment

Issue: High default rates in monthly bill payments		
Theme	**Prevailing respondent sentiment**	**Respondent Characteristics**
Residents view the system as a gift from the government and donors rather than a service requiring continuous financial contributions	*»This project was originally meant for a different village, but because the people there don't appreciate the district leadership, so the leaders decided to put it here because here is the home of their vote. They have a lot of support here.«*	Member of Village Energy Committee
	»This was the product of a political campaign of the leader of this district. You see, this area failed to get electricity for years. Then they decided as a government to give us a microgrid in this village.«	Local government leader
	»Many people do not have enough knowledge of this electricity. Some of them they know that this electricity, they are getting for free because of the campaign. So when you go there to make bill collections, they say, ›No, this is electricity that is for free and we got it as a promise of the campaign.‹«	Member of Village Energy Committee
	»Initially people took this up and we saw a remarkable uptake of the electricity. However, there were some issues...as the project moved forward, the masses lost interest...there is a saying that ›one does not have heart where they have not invested‹... it means if one has invested or felt the pain of something, then they will see its worth. As the project continued, the will of the people left, they no longer took it as something of worth that they respect. Many people failed to pay 5,000Tzshs per month, so this was reduced to 3,000Tzshs, but still people failed to pay. Some even asked us to just take this electricity from their premises.«	Member of Village Energy Committee

Tab. 8b: Case Study Microgrid—Survey themes and prevailing stakeholder sentiment

Issue: Operational mismanagement		
Theme	**Prevailing respondent sentiment**	**Respondent Characteristics**
The VEC lacked the technical expertise to effectively supervise the technician and enforce financial discipline	*»The microgrid operations often failed. The problem is we had no way to know how serious the problem was. The technician tells us that a component is failing, but we do not have the expertise to verify whether that is true or if it is being mismanaged. We did not know so we couldn't query the expenditure.«*	Member of Village Energy Committee
	»Whenever we would attempt to query his expense claims, the technician would disappear for extended periods leaving the microgrid electricity off—as if to punish us. He would only return after repeated begging.«	Member of Village Energy Committee
	»We had nowhere to report him. He was appointed by the District Energy Office but we were never officially introduced to a representative there and no government representative ever came to the village to perform inspections of any kind. It was the technician himself who reported to us he had been appointed to oversee microgrid maintenance.«	Member of Village Energy Committee
	»The village committee exists, but they lack technical expertise. If we want things to run smoothly, we still rely on the technicians and the government office at the district.«	Local government leader
The Energy Committee lacked the expertise to effectively manage the grid	*»Sometimes the energy committee members will collect money but fail to note it down. Then they come back to demand double payment from us. So many of us decided it is better to leave this electricity and live in peace.«*	Microgrid consumer
	»We would just hear that they have gone to the bank, they have withdrawn money... Now they have gone to the bank again. They withdrew this money (microgrid project savings) with a promise that the money would be refunded but it was never refunded. They claimed to have borrowed this money for the development of the village, for renovation of classrooms at the primary and the secondary school, office buildings. That it would be refunded but none of this would be done. As a result, very little of this project would be seen, and the loans would never be repaid. None of this money was repaid. These actions can doom any project.«	Microgrid consumer (former leader)

Tab. 8c: Case Study Microgrid—Survey themes and prevailing stakeholder sentiment

Issue: Disgruntled technical support team		
Theme	**Prevailing respondent sentiment**	**Respondent Characteristics**
The sole technician unhappy with the lack of trust faced from the VEC and discontent from the consumers	*»The key to the microgrid container is kept at the District Energy Office. Those were my instructions... For all repairs I have to pick it and take it back and that place is miles away. Transport is not cheap.«* *»The system is just old. Almost one third of the solar panels have damage. The battery system has needed to be replaced for years. With such a system the frequent failures are not a surprise... These people do not appreciate my struggles. They need a magician.«*	Sole microgrid technician
	»Also for me I have got my own challenges because when the committee calls me, come here to switch or cut off this customer. I am the one they get upset with, the one they quarrel with. To them, I am the enemy. Even I fear for my life, when I arrive there, I don't eat at people's houses because I may be poisoned! The committee makes no effort to understand my problems. Maybe I was sick and unable to work, it becomes a big problem.«	Sole microgrid technician
Issue: Unregulated power consumption patterns		
Power wastage, unregulated consumption, and frequent outages	*»The issue is that flat rate tariff that doesn't count actual consumption. Because there is no difference in your fees whatever you consume, people have no incentive to conserve energy. They just consume without care.«*	Member of Village Energy Committee
	»So much electricity is wasted. You see many security lights on during the day. There is no effort to conserve power. That is why the system is off at night when we most need it because it was depleted during the day by such people.«	Microgrid consumer
	»We were only asked whether we wanted to be connected or not. I was never taught about saving power. I know my obligation is to pay the monthly fees and this I do religiously. If we all pay the fees the system should not go off so much.«	Microgrid consumer

Tab. 8d: Case Study Microgrid—Survey themes and prevailing stakeholder sentiment

Issue: Financial leakages		
Theme	**Prevailing respondent sentiment**	**Respondent Characteristics**
Collected funds diverted by the Village Energy Committee from microgrid maintenance to more urgent community initiatives	*»We had a bit of saving on the account but as you know with money, the VEC decided it was too idle and appropriated it on loan to more pressing community initiatives like the village school and the health centre. Of course it was never repaid. And now when we needed to spend on maintenance, there was nothing.«*	Member of Village Energy Committee
	»It was not a unanimous VEC decision (to divert the money) but the dissenters were in the minority and they received immense pressure. The sentiment was they were blocking community development. In the end there was nothing we could do.«	Member of Village Energy Committee
	»The community was not involved in the decision (to divert collected funds)—the committee made the decision in interest of the community.«	Member of Village Energy Committee
	»We were never consulted on VEC decisions. Those are our leaders. We only pay because we are told to pay. We don't know how they decide the costs or how the system is maintained. I do not remember them ever consulting us on anything. They are supposed to make the decisions.«	Microgrid consumer

4.2.3.3 Navigating Power and Technical Inequality

Our thematic analysis of the interview data identified three key areas impacting microgrid sustainability: (1) Power asymmetries in governance; (2) Technical knowledge gaps and dependency; and (3) Weak enforcement and accountability mechanisms.

4.2.3.3.1 *Power Imbalances Reinforced by Technical Dependence in the Microgrid Community*

Interviews with VEC members and local government representatives highlight that decision-making within the microgrid is heavily centralized, with technicians and external government agencies exercising disproportionate control. Both VEC members interviewed expressed frustration at their limited authority over system operations, as key technical and financial decisions are made by the government-appointed technician without their meaningful input. Similarly, local government representatives acknowledged the dual governance structure, where both the District Energy Office and the VEC claim supervisory authority. Consumers, in turn, perceive the microgrid as externally controlled, leading to low engagement and compliance with tariff payments.

In microgrids like the one presented here, technical knowledge for maintenance, power distribution, and overall system health is highly specialized. However, this knowledge is not evenly distributed, creating an expanding gap between those who operate the system (technicians) and those who are supposed to govern it (community committees and users). The community's low levels of education exacerbate the knowledge gap, limiting local capacity to acquire and apply these skills. This growing knowledge gap increases dependence on technicians, even for minor operational decisions, reinforcing *technocratic control* over governance. As a result, community committees struggle to challenge technical decisions, often accepting operational inefficiencies or financial mismanagement without question. Additionally, limited user awareness of system constraints raises the risk of overuse, non-compliance with energy-saving guidelines, and eventual system failures. The result is a self-reinforcing hierarchy where the VEC remains dependent on technician(s), who in turn operate with minimal accountability. This governance structure directly contradicts STS principles, which emphasize participatory decision-making and distributed knowledge access as essential for sustainable socio-technical systems (Pitt et al., 2018). Without these elements, the system became resistant to governance reforms, leaving non-technical stakeholders disempowered and further weakening its long-term sustainability.

4.2.3.3.2 Power Asymmetries Undermine Collective Governance

Results have shown the community microgrid to experience conflicting dual governance structures, where both a government energy office and a community-led committee claim supervisory authority. This institutional ambiguity created overlapping responsibilities, weakened enforcement mechanisms, and left power centres in terms of the technician unchecked, ultimately leading to poor rule compliance and financial mismanagement. Moreover, the exclusion of local stakeholders in terms of the consumer community from decision-making has led to governance failures that have weakened the sustainability of the microgrid.

The interplay between hierarchical governance and disparities in technical capacity creates a self-reinforcing cycle that undermines microgrid sustainability. Fig. 44 illustrates the key dynamics. Persistent financial mismanagement and weak oversight arising from the community's exclusion reduce the system's sustainability, leading to more frequent breakdowns, growing consumer distrust, and increased non-compliance. Without governance reforms, technician dominance continues to hinder participatory decision-making, entrenching top-down control and limiting community involvement in system management.

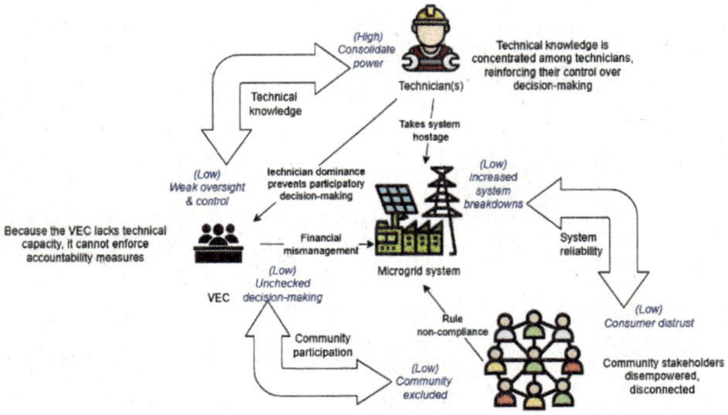

Fig. 44: Reinforcing cycle of hierarchical governance and technical dependence in microgrid management.

4.2.3.5 A Democratized Technical Training Framework for Microgrid Governance

From CPR theory, sustainable management of shared resources requires collective governance, participation in rule enforcement, and equitable decision-making structures (Ostrom, 1990). When only a small subset of stakeholders—such as technicians—possess technical expertise, the community remains passive in microgrid oversight, reinforcing financial mismanagement, weak enforcement, and system inefficiencies. To align technical training with CPR principles, a revised microgrid training model must focus on distributing technical knowledge beyond technicians to include Village Electricity Committees (VECs) and select community representatives. By incorporating co-learning structures where knowledge flows both from technicians to the community and vice versa, a CPR-based training framework transforms microgrid management from a technician-controlled system to a collective governance structure.

From an STS perspective, technology and governance are deeply intertwined; microgrid sustainability is not just a technical challenge but also a social and institutional issue (Bijker et al., 1987). Traditional technical training models assume that knowledge should be concentrated in a select group of experts—a model that reinforces hierarchical power structures rather than fostering distributed governance. To create an STS-aligned training model, technical training must be redesigned to break down knowledge asymmetries by making system performance data accessible and interpretable to non-technical stakeholders; to shift from expert-led training to participatory co-design models, where VECs and community members actively engage in troubleshooting and decision-making processes; and to encourage peer-learning systems, where trained community representatives become local knowledge multipliers, reducing reliance on external agencies. These principles emphasize that successful microgrid governance requires not only technical expertise but also institutional structures that allow communities to apply this knowledge meaningfully in decision-making.

From the two theoretical frameworks, we propose a democratized technical training framework (DTTF) that integrates technical and social knowledge for inclusive decision-making. This CPR-STS-aligned training framework is built on three core pillars: (1) decentralization of technical knowledge ensuring that technical expertise is distributed beyond technicians to include local governing bodies and end-users; (2) participatory and inclusive governance structures that shift decision-making power from technicians and external actors to inclusive, community-led mechanisms; and (3) transparent oversight mechanisms including financial and technical accountability structures that enable all stakeholders to monitor system operations and performance. The DTTF aims to restructure microgrid training to both enhance technical skills and redistribute decision-making power. Fig. 45 summarizes the core pillars of the DTTF.

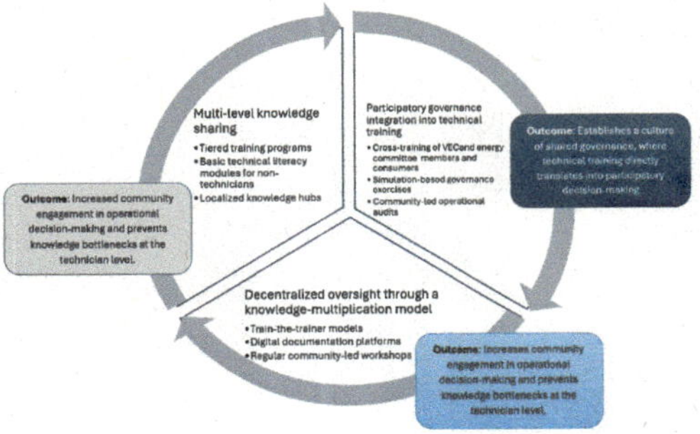

Fig. 45: The Democratized Technical Training Framework (DTTF).

The DTTF proposes a restructuring of microgrid training by addressing governance asymmetries reinforced by technical dependence. Technical Knowledge Decentralization (core pillar I) provides multi-level training for VECs and community members, ensuring broader system oversight; Participatory Governance Integration (core pillar II) aligns technical training with governance roles through cross-training in oversight and simulation-based decision-making exercises; and Decentralized Oversight (core pillar III) promotes train-the-trainer models, community-led operational audits, and transparency tools to prevent technician monopoly over system knowledge. By integrating CPR theory's focus on collective governance and STS theory's emphasis on knowledge co-design, the DTTF framework is expected to equip local stakeholders with basic system monitoring and financial auditing skills, strengthen VEC authority in financial and operational decisions, reduce technical dependence, and increase accountability. Further, by breaking down technical knowledge silos and embedding participatory governance mechanisms into training, DTTF enables greater local ownership, financial accountability, and long-term sustainability of community microgrids.

4.2.3.6 Conclusion

The results confirm that hierarchical governance structures and technical capacity disparities interact with each other, thereby influencing microgrid operations and long-term sustainability. While CPR theory highlights how power asymmetries weaken collective governance, STS theory explains why technical dependency reinforces these governance failures. A sustainable microgrid governance model must (1) distribute technical knowledge beyond technicians to empower local decision-making, (2) reform governance structures to ensure participatory decision-making rather than technician-dominated control, and (3) strengthen enforcement mechanisms through financial transparency and operational oversight.

By addressing both the governance and technical capacity dimensions simultaneously, microgrids can transition from technocratic, top-down systems to inclusive, community-driven energy solutions, ensuring long-term financial viability and operational efficiency.

4.2.4 How Ownership Matters for Sustainable Implementation of Microgrids

Irene F. Ngoti

4.2.4.1 Introduction

Microgrid sustainability is a growing concern for people in rural-developing nations given the inaccessibility of the national energy grid. Independence from the centralized energy system in these areas is due to several reasons such as the existence of a scattered settlement in rural areas, which makes it expensive for the national utility to supply electricity, and the financial difficulties and high poverty levels facing the majority of rural dwellers, which in turn rarely provide investment returns for the energy investors. For such reasons, decentralized energy systems through microgrids emerge as among the best alternatives for rural communities to experience almost what urban people have in terms of electricity supply (IEA, IRENA, UNSD, World Bank, WHO, 2022; Tenenbaum et al., 2014). This does not imply that all microgrids can accommodate the scattered and poor settlement in rural areas and that microgrid investors can receive returns on investment following their decision to invest in rural areas; rather, there exist different microgrid ownership models with various technologies, goals, financial capacity for different rural populations. The following are the common microgrid models that exist in developing countries and specifically in SSA: private sector model, community-based model, utility-based model, NGO model, and hybrid ownership model (Odarno et al., 2017). The institutional environment (institutional arrangement and regulations) for a particular country or locality often dictates the microgrid ownership model for energy supply in rural communities. The governments are therefore responsible for creating suitable institutional environments that include legal status and security, subsidies (for attracting investors), and guarantees for continuity of services for microgrid ownership models (Pérez-Arriaga, 2022). Investors and respective microgrid beneficiaries are important actors responsible for a particular microgrid ownership model. We therefore aim to examine the significance of ownership in relation to sustainable implementation of microgrids, taking a case study from Tanzania. We first go through different microgrid ownership models before going into a detailed analysis of the case study at hand.

4.2.4.2 Microgrid Ownership Models

Private sector ownership model. Under this model, a private investor is solely responsible for the microgrid installation, operation, maintenance, and management. The funding sources for investors are mainly private equities, commercial or concessional loans, and even government grants (Energypedia, n.d.). Countries with supportive microgrid regulatory frameworks and policies often attract private ownership models. This model also has a higher probability of accessing credit compared to the community-based ownership model because of the well-defined business model for the majority of investors. Profit-making is usually the main focus under this model. To attain profit and finance loans (if available) amid rural communities with low-income earners, the business model for this ownership model accommodates an anchor load approach (in addition to business and normal customers) and/or a key-maker business operation model. On the customers' side, this model often charges higher tariffs than other ownership models, to recover investment costs and earn some return on their investments. With good business models, this ownership structure provides reliable and sustainable electricity to its users.

Community-based ownership model. In this model, local communities own, manage, operate, and maintain microgrids. In developing countries (e.g., East African countries), village energy committees (VECs) who represent respective energy beneficiaries in rural communities are usually responsible for managing the majority of microgrids under this model. Because of the limited technical capacity in the majority of rural areas, microgrid installation is carried out by external parties (e.g., developers, outside organizations). Funding is mainly sourced from grants and government, with very few solely communally funded microgrids. This model is also characterized by low tariffs being charged to its users, and the probability of success comes with existing programmes that support productive end-use of electricity together with local stakeholders' acceptance, active engagement, and participation. A good revenue collection model, community engagement, and participation are additional factors that influence the reliability and sustainability of this model.

Utility-based ownership model. Government-owned utilities under this model manage, operate, and maintain microgrids. Since the microgrids in this ownership model are state-owned, they receive subsidies to charge affordable tariffs to users. Governments with national electrification plans/programmes tend to support this particular model, which in turn enhances sustainability.

Non-governmental ownership model. In this ownership model, microgrids are owned, managed, operated, and maintained by non-governmental organizations such as religious organizations. The main source of funding comes from donors. Users under this model are charged low tariffs because the model's focus is not profit, and it highly subsidizes tariffs. Microgrids under this ownership model are usually sustainable despite the subsidized tariffs to their users because of funding sources (grants) and the supportive productive use activities that generate income to finance the microgrids' operations.

Hybrid ownership model. This model combines two or more of the preceding models. It is common when one of the actors from the preceding models lacks the capacity or expertise for either microgrid design, installation, operation, management, or maintenance. This model may face conflict among stakeholders if there is no strong microgrid regulatory framework; otherwise it can be an effective approach beneficial to both owners and electricity end-users.

4.2.4.3 The Case of a Community Microgrid Project in Tanzania

4.2.4.3.1 *Project overview*

The chosen microgrid project is among 14 community-based containerized solar microgrids constructed across 10 villages in Tanzania. The government, through the Ministry of Energy and Minerals, for the first time took the initiative to elevate community-based energy projects to implement its rural electrification plan and policy. The 14 microgrid projects were targeted as models for more microgrid development in other rural areas of Tanzania in the future. Interviews conducted with different stakeholders in the village were able to provide rich information in this section. More detailed project information on the microgrid can be found in Ngoti (2024a; 2024b). Fig. 46 below gives an overview of its features.

4.5.4.3.2 *The ownership structure of the community microgrid*

The microgrid follows a community-based ownership model that has gone through two-phase transitions because of a change in management structure.

Initial microgrid ownership and development phase. This was the first phase where the Tanzanian government, via the Ministry of Energy and Minerals, owned the microgrid. The project was funded with a soft loan

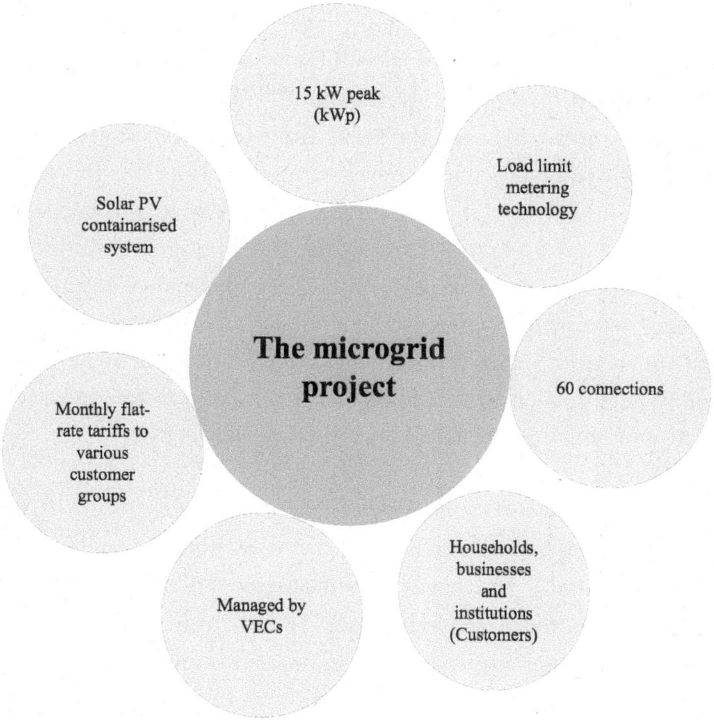

Fig. 46: Features of the Tanzanian microgrid (Ngoti, 2024a; 2024b).

from the Austrian government (Odarno et al., 2017). The ministry contracted an electrical company to install, operate, and manage the microgrid in the first three years. It was the company that conducted microgrid maintenance while the role of tariff collection was left to the VEC. The land for microgrid construction was freely given by the respective village following the agreement of microgrid installation and operations in the village. In this period, the (legal) ownership solely remained with the government of Tanzania through the Ministry of Energy and Minerals.

Transition to community ownership phase. In this phase, the electrical company stopped managing and operating the microgrid because its contract had expired. There was then a transfer of roles that were performed by the company to the hands of the VEC and village office that represented microgrid users. There was no legal transfer of ownership or official transfer of ownership from the government to the microgrid users. It was instead agreed by word of mouth between the two parties that there would be an official transfer of ownership, which never happened. The VEC therefore took full control of the financial management, maintenance, and decision-making on all matters relating to the microgrid. The microgrid community users therefore played a direct role in the microgrid governance despite the absence of legal ownership transfer to them. The nature of ownership that existed in the microgrid is what is referred to as »psychological ownership« or »sense of ownership« which is defined as a psychological state in which people (microgrid users) feel that the microgrid project is theirs (Pierce et al., 2001).

Linking ownership to sustainable management of the microgrid—User perspectives. Ownership status provides the owners with different benefits that have positive impacts on what is owned. Sustainable implementation or management of microgrid is vital as users feel assured by reliable electricity services. The ownership status of a microgrid is an essential factor that is neglected in the literature, but it has an impact on the microgrid's sustainability. In this subsection, we consider the significance of microgrid ownership for sustainable microgrid management by examining the microgrid users' perspectives.

Promoting security for microgrid assets. The existence of a sense of ownership among users promotes the urge to protect the microgrid assets to harness the microgrid benefits sustainably. Users in the community, regardless of their positions, all engaged in protecting the microgrids. They knew the project was theirs; even if they had no legal ownership over it, it still belonged to them. One of the village leaders had the following to say: »*this is a project that the villagers know is their property... therefore, in case of anything, all the villagers know it is their property. If you attempt to do anything, the villagers are vigilant. Generally, the microgrid protection is in the hands of the villagers and they have* realized *that it belongs to them*«.

Autonomous use of tariffs funds. Microgrid ownership has many benefits associated with the autonomous use of tariffs collected. With no external interference on the use of tariffs, the fund is available when needed for maintenance and repairs, paying technicians, and buying spare parts. This improves service reliability and eliminates elements of bureaucratic delays that involve the need for external approvals when for instance there is a need for urgent maintenance. One user explained: »... *and later they realized that this project is really ›our‹ property because the money collected is not going to be taken by* anyone, *the government, or the company; they manage the revenues and expenditure by themselves; it dawned on them that it is ›their‹ property*«.

Tenacious microgrid financing. A clear ownership structure plays a crucial role in ensuring that the tariff collected is directed to a proper channel for microgrid sustainability. In the Tanzanian village community, microgrid users stood firm to make sure that the maintenance fund is not empty despite the ups and downs in microgrid operations, because of a high sense of ownership among the users. As one of the users explained, »... *since the beginning of the project, we are contributing, contributing, contributing, to facilitate our project, that is, we are financing it so that it becomes sustainable.*« This statement demonstrates that a sense of ownership pushes users to uphold the microgrid's sustainability. This spirit in the village facilitated the long-term viability of their microgrid which further attracted donors who also seek stability in microgrid operations.

4.2.4.4 Conclusion

Various microgrid ownership models have both advantages and disadvantages, which may in turn affect microgrid sustainability objectives. However, the project's operating environment determines which microgrid ownership structure fits best (Liu & Starke, 2025). The following considerations are vital when deciding which ownership model to adopt so as to ensure a sustainable microgrid implementation:

Technical and managerial expertise: Different ownership models bring diverse technical and managerial expertise to the project. Private companies through private ownership may have more experience in operation and maintenance because of their stable funding base and good business model, while community-based ownership may excel in understanding local needs and preferences. Their effects on sustainable microgrid implementation may therefore be the same but applied from different angles, i.e., the private ownership model has its strengths on the technical side while the community-based model can use community engagement and participation in decision-making and general microgrid management.

Tariffs and pricing: The ownership structure can influence electricity tariffs and pricing. Community or public ownership may prioritize affordability and set tariffs that take into account the local population's economic capacity while private ownership may focus on profit maximization and thus implement higher tariffs for its users. These are two different approaches, but they may arrive at the same goal of sustainable microgrid implementation.

Project funding sources and costs: Community and non-governmental ownership models are mostly financed through grants and thus tend to charge low tariffs to their customers. Community microgrids are commonly developed in remote areas where it is not cost-effective for public or private investors to maintain them, e.g., in areas where there are no anchor customers. Private investors can choose where to invest to finance their loans (if any) and have investment returns. These distinct modalities imply that each model is for a specific group of customers and a mere misplacement may have negative implications on microgrid sustainable implementation.

Government policy and regulatory framework: A regulatory framework that is clear, transparent, consistent, and enforceable attracts investors. Regulations that are not cumbersome and that protect investors are good for investors within various microgrid ownership models. Different ownership models may face diverse requirements or incentives from the government. All these features have different implications for microgrid sustainable implementation.

4.2.5 Sustaining Off-Grid Energy through Informal Governance: Network Evidence from a Community Microgrid in Tanzania

Lillian Donna Namujju, Burkhard Hehenkamp

4.2.5.1 Introduction

Decentralized energy solutions, particularly off-grid community microgrids, have become critical in addressing rural electrification challenges in SSA. While microgrids are often framed within formal regulatory and policy frameworks, in practice, many operate under weak formal oversight, relying on community-led governance structures (Wentz & Pappalardo, 2016; Kirubi, 2009). These informal institutions play an essential role in ensuring system continuity, enforcing compliance, and mediating energy-related conflicts—yet their role remains underexplored in energy governance research.

This study examines a community microgrid in rural Tanzania as a case study of how off-grid energy systems are sustained in the absence of strong formal oversight. Established through a public-private partnership and later transitioned to community ownership under the Village Electricity Committee (VEC), the microgrid has faced persistent governance challenges, including weak institutional accountability, poor tariff compliance, and technical dependency (Ngoti, 2024a, 2024b). Using a network-based approach, this research investigates how informal institutions—such as the technician, traditional leaders, and local brokers—emerge as key actors in maintaining system operations, enforcing local energy rules, and mediating disputes. It specifically asks: (1) How do informal institutions sustain governance where formal oversight is weak or absent? (2) What is the extent of their authority and legitimacy in microgrid management? And (3) How do non-regulatory enforcement mechanisms influence the long-term viability of off-grid systems? Through social network analysis, the study maps perceived power relationships and governance flows, offering empirical insights into the dynamics of informal authority and its implications for the sustainability of decentralized energy systems.

Existing studies on microgrid governance often emphasize formal regulatory structures and technical challenges but pay limited attention to the role of informal governance mechanisms (Derks & Romijn, 2019; Namujju et al., 2023). Yet, in many off-grid communities, energy governance is de facto negotiated through informal institutions, social norms, and community-based decision-making processes (Eklund et al., 2023). The absence of research on how these informal structures operate autonomously, rather than as adjuncts to formal institutions, leaves a critical gap in understanding decentralized energy governance in weak regulatory environments. The gap leads to overlooked power asymmetries where informal actors accumulate decision-making authority without accountability as well as institutional fragmentation, where formal and informal governance structures operate in parallel but lack coordination.

4.2.5.2 Institutional Concepts and Network Approach

4.2.5.2.1 *Formal vs. Informal Institutions*

New Institutional Economics (NIE) provides a useful framework for understanding how institutions—both formal and informal—shape economic and governance outcomes. North (1990) defines institutions as the rules of the game in a society, comprising both formal structures (laws, policies, regulations) and informal mechanisms (social norms, customs, and practices) that influence behaviour. In the context of off-grid energy governance, this distinction is critical in explaining why and how decentralized energy systems function in the absence of strong formal oversight.

Formal Institutions are the codified rules and enforcement mechanisms including national energy policies and rural electrification strategies advanced by regulatory bodies responsible for licensing, tariff-setting, and oversight. Informal Institutions, on the other hand, operate outside formal regulatory structures but play a critical role in governance, like social norms governing energy access and consumption behaviour, traditional leadership roles in resolving disputes over microgrid management, etc. In many rural off-grid energy markets, these informal institutions emerge not as substitutes but as primary governance mechanisms due to the weak enforcement of formal rules.

In weak regulatory environments, formal regulations are either poorly enforced or entirely absent, leading to de facto governance by informal institutions. Under these conditions, informal governance structures compensate for formal institutional gaps so that, instead of state-led enforcement, peer accountability, local social norms, and traditional leadership influence consumer behaviour; informal mediation is adopted for dispute resolution rather than formal legal or regulatory processes, and local actors and informal brokers take on decision-making roles in the absence or weakness of formal oversight.

4.2.5.2.2 *Social Network Analysis in Energy Governance Research*

Social Network Analysis (SNA) offers a powerful tool for examining how governance actors interact, influence decision-making, and shape institutional structures within off-grid energy systems. SNA enables the identification of which institutions or individuals wield the most influence in governance, regardless of their formal authority, the degree of centralization or decentralization in decision-making, and the presence of gaps (structural holes) or redundancies in governance networks. Applying SNA to the community's microgrid governance moves beyond the traditional formal/informal dichotomy, providing deeper insights into who truly controls decisions, enforces rules, and oversees system operations in practice.

4.2.5.3 Methodology

The study follows a qualitative case study approach combined with SNA to assess how governance actors—both formal and informal—shape microgrid sustainability. The research focuses on mapping governance relationships to identify how actors interact; assessing power dynamics and decision-making structures using SNA measures; and identifying governance gaps (structural holes) and redundancies (overlapping roles). A mixed-methods approach is used, integrating semi-structured interviews, social network surveys, and secondary data analysis.

4.2.5.3.1 *Study Context*

To explore governance relationships in the community microgrid, this study identifies and maps the key actors involved in system oversight, decision-making, and service delivery. The governance network comprises both formal and informal institutions. The Village Electricity Committee (VEC) functions as the primary community-based entity responsible for financial and administrative management. The technician, appointed by the District Energy Office (DEO), is tasked with system maintenance and technical operations. Local community leaders, whether traditional or elected, hold social authority and are frequently consulted on energy-related matters. Local energy brokers act as informal intermediaries, assisting consumers with troubleshooting and facilitating informal payment arrangements, despite lacking formal appointments. The DEO itself holds a nominal supervisory role as a formal government agency. The electrical company, a private developer, was responsible for the initial installation of the microgrid under a public-private partnership but is no longer involved in day-to-day governance. The Rural Energy Agency (REA) is the national-level regulatory and planning body for rural electrification. Lastly, consumers represent the end-users of the microgrid and constitute the majority of actors surveyed regarding perceptions of governance influence.

4.2.5.3.2 *Data Collection*

To explore governance relationships, in-depth interviews were conducted with informal actors such as microgrid technicians, VEC, local community leaders, local energy brokers, and household consumers. Interviews focused on decision-making roles, enforcement mechanisms, and operational challenges, ensuring a detailed understanding of who governs what aspects of the microgrid. No interviews were possible with formal actors such as DEO, REA, or the electrical company. Information on their roles and challenges was inferred from informal actor interviews and secondary data. Project documents, tariff records, and meeting minutes (where available) were analysed to supplement interview and survey findings, helping to contextualize governance interactions.

A governance network survey was conducted to systematically map decision-making relationships among key microgrid actors. Respondents were asked to identify who holds primary decision-making authority over six key governance functions: (1) Repairs & maintenance—who determines when and how system repairs are conducted? (2) Tariffs—who sets or influences the electricity pricing structure? (3) Collections—who is responsible for enforcing payment compliance? (4) Connections & disconnections—who decides on household or business connections and disconnections due to non-payment or other issues? (5) Disputes—who mediates conflicts related to electricity access, payments, or service disruptions? (6) Allocation of microgrid revenues—who controls financial decisions regarding revenue distribution or reinvestment? (7) Overall microgrid control—who has overall control over the microgrid?

4.2.5.4 Data Synthesis and Analysis

The responses were used to construct a governance network map in R, illustrating relational influence—that is, who is perceived to hold decision-making authority across specific governance functions (e.g., tariffs, maintenance, disconnections). By connecting *perceivers* (those identifying authority) to the actors they perceive as decision-makers, the study generated a directed governance network, where each edge represents a perceived authority link. The direction of each edge flows from the perceiver to the perceived power holder, while the edge weight reflects the number of respondents who identified that actor as influential. This approach offers a visual and analytical representation of power dynamics, revealing decision-making hierarchies, institutional overlaps, and structural gaps in governance and enforcement. To further interpret these relationships, the study applies an SNA measure—weighted in-degree centrality—which measures not only how many actors attribute decision-making authority to a given actor, but also how frequently that actor is named across multiple governance functions—providing a more nuanced view of perceived power concentration. Additionally, the study generated a weighted adjacency matrix based on these perceptions, which was used as input for hierarchical clustering analysis. This allowed for the identification of governance clusters, revealing structural groupings of core decision-makers versus peripheral or inactive institutions within the governance network.

4.2.5.5 Results and Discussion

In this section, the findings from the SNA of governance perceptions within the community microgrid are presented. Drawing on data from community respondents and key stakeholders, the analysis reveals how different actors are perceived to hold decision-making authority across critical governance functions such as tariff-setting, maintenance, disconnections, collections, and dispute resolution. Using a directed and weighted governance network, the results highlight patterns of influence, institutional clustering, and the dominance of informal structures in the absence of strong formal oversight. Key insights are organized around actor-level influence (using weighted in-degree centrality), the structural layout of the governance network (via hierarchical clustering), and visual mapping of power dynamics through heatmaps and network diagrams. Together, these findings illustrate the concentration of governance authority, the limited role of formal institutions, and the centrality of local actors in shaping microgrid operations.

4.2.5.5.1 *Stakeholder Perceptions of Governance Influence in the Tanzanian Community Microgrid*

Fig. 47 shows the perceived distribution of actor influence across the different governance functions in the microgrid. The technician holds a dominant role in system operations, being perceived as responsible for 94 % of repairs & maintenance and 100 % of connections and disconnections. This indicates a monopoly over technical and access-related decisions. Additionally, 19 % of respondents associate the technician with tariff-setting, suggesting that technical constraints may be leveraged to influence pricing decisions. The VEC leads in governance of tariffs (75 %), collections (81 %), dispute resolution (81 %), and revenue allocation (75 %), firmly establishing its control over financial and administrative management. Local brokers are notably involved in collections (50 %) and repairs & maintenance (13 %), implying that some consumers bypass the VEC in favor of intermediaries for tariff enforcement or seek informal troubleshooting arrangements within the community. Local community leaders are perceived to influence disputes (38 %) and revenue allocation (63 %), highlighting their informal authority in local governance. Their involvement in tariffs (25 %) further suggests occasional consultation or influence in pricing decisions. The DEO plays a limited role, with perceived involvement ranging from 13–19 % and demonstrates minimal engagement in active decision-making. Meanwhile, the electrical company and REA are entirely peripheral, indicating that their roles in governance are largely symbolic or dormant. This pattern points to weak formal oversight and a governance vacuum that is filled by local and informal actors.

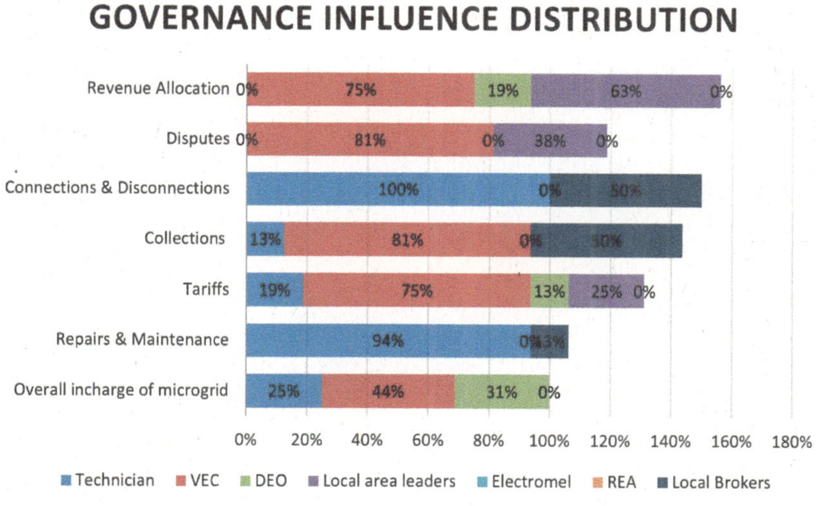

Fig. 47: Governance influence distribution in the community microgrid.

The results reveal critical insights into the microgrid governance. The heavy reliance on the technician for technical operations and access control raises concerns about unilateral decision-making and limited accountability. While the VEC holds primary financial authority, the involvement of local brokers in collections points to gaps in formal enforcement mechanisms. Local community leaders play a significant role in dispute resolution and revenue allocation, underscoring the influence of informal governance structures. Notably, the absence of formal oversight actors such as the DEO and the REA in key areas like enforcement, financial management, and conflict resolution highlights a lack of regulatory accountability. As a result, the microgrid's governance is shaped predominantly by local and informal actors, rather than structured institutional oversight.

4.2.5.5.2 Structural Layout of the Microgrid Governance Network

Fig. 48 and Fig. 49 illustrate how governance influence is concentrated and distributed among different actors in the microgrid governance structure.

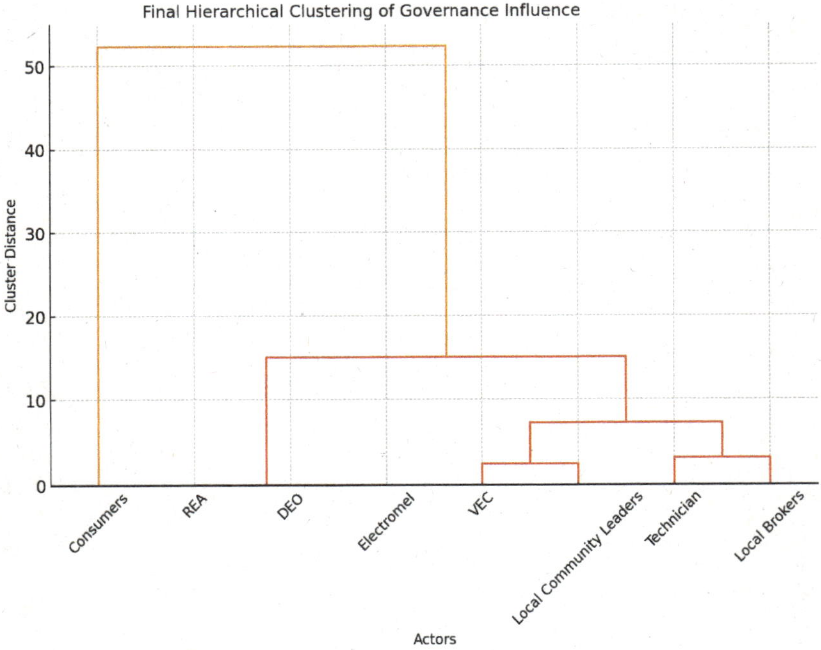

Fig. 48: Hierarchical clustering of governance influence.

Fig. 48 reveals two distinct governance clusters. Cluster 1 consists of the core microgrid decision-makers—the VEC, technician, local community leaders, and local brokers—who are highly interconnected and hold the majority of governance influence across key functions. In contrast, Cluster 2 includes the electrical company, REA, DEO, and consumers, who remain on the periphery of the network with minimal or no involvement in decision-making. Notably, consumers are completely disconnected from governance processes, reinforcing their passive role within the microgrid's institutional structure.

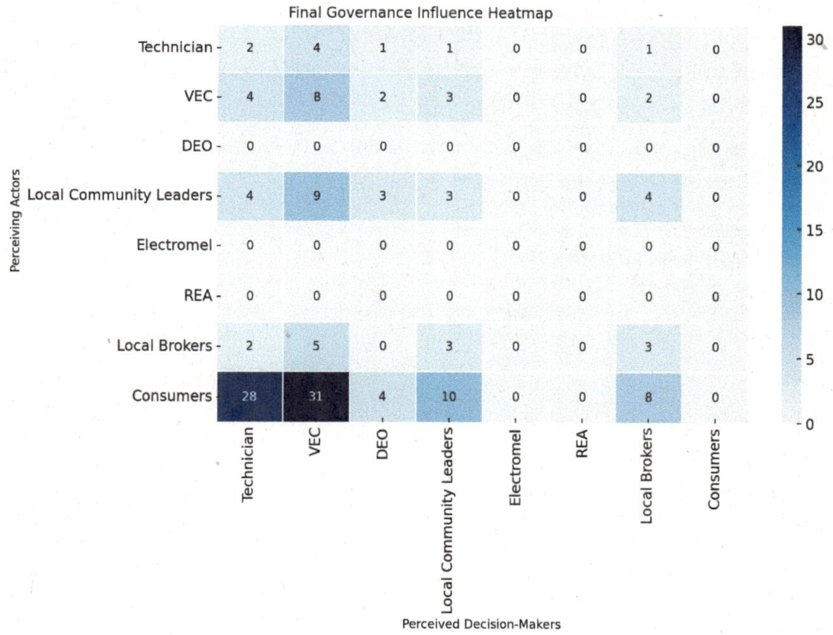

Fig. 49: Governance influence heatmap.

As shown in the heatmap in Fig. 49, the VEC and the technician emerge as the dominant governance actors, receiving 31 and 28 attributions of authority from consumers, respectively. Local community leaders (10 attributions) and local brokers (8) serve as secondary power holders, particularly in areas related to financial enforcement and informal governance support. In contrast, the DEO, REA, and the electrical company are largely absent from the governance landscape, highlighting the limited role of formal institutions and a lack of effective oversight. While consumers clearly recognize the existing power structures, they are not perceived by any other actors as holding decision-making authority, underscoring their exclusion from the governance process.

4.2.5.5.3 Actor-level Influence in the Community Microgrid

Analysis of weighted in-degree centrality provides a clear picture of how governance authority is distributed across actors in the community microgrid. This metric captures not just the number of times an actor is perceived to hold decision-making power, but also the frequency and strength of those perceptions, offering a robust measure of influence. Tab. 9 shows a breakdown of in-degree centrality by actor.

Tab. 9: Actor-level Influence

Actor	Weighted In-Degree Centrality	Normalized In-Degree Centrality
VEC	57	1.000
Technician	40	0.702
Local Community Leaders	20	0.351
Local Brokers	18	0.316
DEO	10	0.175
Electrical company	0	0.000
REA	0	0.000
Consumers	0	0.000

The VEC stands out as the most influential actor, with a normalized in-degree centrality of 1.00, indicating that it is the most consistently perceived authority across the governance network. The 0.30 gap between the VEC and the technician (0.70) is significant on a normalized scale, underscoring the VEC's disproportionate dominance, particularly in financial governance. While the technician also holds considerable influence, especially in operational matters, the substantial gap between the technician and the next tier of actors—local community leaders (0.35) and local brokers (0.32)—further highlights the divide between formal authority and informal, supporting roles. The steep decline in perceived influence from these actors to the DEO (0.17), and the complete absence of the electrical company, REA, and consumers (0.00), reflects structural marginalization and the absence of effective formal oversight. This sharp gradient in influence reinforces the characterization of the community's governance structure as highly localized and informally driven, rather than balanced across a diverse set of institutional actors.

Overall, these results point to a highly centralized governance structure, concentrated in the hands of a few local actors, with informal mechanisms filling the void left by weak formal oversight and absent regulatory institutions. The normalized scores provide a clearer view of relative influence and the structural imbalances that define governance in the community's microgrid.

4.2.5.6 Conclusion and Policy Implications

The community microgrid case highlights the need to rethink governance models for off-grid energy systems, especially in rural settings with limited regulatory oversight. While local actors like the VEC and technician play essential roles, their unchecked authority reveals risks tied to weak accountability. To address these governance imbalances, formal institutions such as the DEO and REA must be strengthened and actively involved in monitoring and oversight. Their absence leaves room for informal power structures to dominate without accountability. Clear role definitions between technicians, committees, and supervisors are necessary to guard against the consolidation of power in the hands of a few individuals or committees. Moreover, where consumers are passive with no perceived decision-making power, policy should promote meaningful consumer participation in decision-making processes—such as tariffs, revenue use, and dispute resolution—transforming end-users from passive recipients to co-governors. Finally, since local community leaders and brokers already perform key enforcement and mediation roles, policy approaches should consider formalizing or legitimizing their contributions through structured participation or delegated responsibilities.

The network structure exhibits significant power asymmetries, with consumers entirely excluded from decision-making processes and formal institutions failing to provide meaningful oversight. These findings highlight a fundamental governance gap in off-grid systems operating under weak regulatory environments, where informal institutions fill the void left by disengaged formal actors. Strengthening microgrid sustainability will require re-balancing this structure by enhancing the role of formal institutions, improving transparency and accountability, and expanding participatory mechanisms that include consumers and community stakeholders as active agents in governance.

4.2.6 Conclusion—Sustainability and Performance Outcomes in Microgrid Operations

Burkhard Hehenkamp

In this section, we have investigated the role of socio-cultural and institutional factors in the sustainable functioning of microgrids. This allowed us to identify important parameters that promote or hinder the sustainability of a microgrid. Summarizing our insights, this conclusion serves to derive some recommendations for microgrid governance.

In Section 4.2.2, we found out that institutional factors—compared to the socio-economic factors under investigation—play a major role in driving households' willingness to pay (WTP) for microgrid services. Promoting prepaid billing, enhancing fairness perceptions, and reducing barriers to digital payment represent actionable strategies.

The interplay of institutional and socio-cultural factors was the focus of Section 4.2.3. There we saw that power asymmetries based on technical dependency weaken collective governance. Sustainable implementation must therefore: (1) distribute technical knowledge beyond technicians to empower local decision-making; (2) reform governance structures to ensure participatory decision-making; and (3) strengthen enforcement mechanisms through financial and operational oversight.

As to the optimal ownership structure, we argued in Section 4.2.4 that each model of ownership has its advantages and disadvantages. As to technical and managerial expertise, private or public ownership typically involves more experience in operation and maintenance, while community-based ownership has the advantage of having a better understanding of local needs and preferences. Community and public ownership may prioritize affordability and involve lower tariffs, whereas private ownership may focus on profit maximization and thus lead to higher tariffs. In the end, the specific environment of the project determines which model of ownership fits best.

Finally, we have scrutinized a community microgrid in Section 4.2.5 to disclose adverse factors that compromise the sustainable functioning of microgrids. We found governance imbalances and informal structures that had emerged, assuming and replacing the role of formal institutions, which were not enforced effectively. These informal structures involved power asymmetries and turned out to be detrimental for the sustainability of the microgrid.

On a more general level, we have learned that the institutional and regulatory framework that governs the microgrid should:

- rely on active participation of the community, creating at least a feeling of ownership;
- involve a well-balanced distribution of power, responsibility, and accountability;
- be clear, transparent, and enforced.

Only then can a sustainable functioning of the microgrid be expected.

4.2.7 References

Abdullah, S., & Jeanty, P. W. (2011). Willingness to pay for renewable energy: evidence from a contingent valuation survey in Kenya. *Renewable and Sustainable Energy Reviews, 15*(6), 29742983.

Ahlborg, H., & Hammar, L. (2014). Drivers and barriers to rural electrification in Tanzania and Mozambique—Grid-extension, off-grid, and renewable energy technologies. *Renewable Energy, 61*, 117–124.

Akinyele, D., Belikov, J., & Levron, Y. (2018). Challenges of microgrids in remote communities: a STEEP model application. *Energies, 11*(2), 432.

Alstone, P., Gershenson, D., & Kammen, D. M. (2015). Decentralized energy systems for clean electricity access. *Nature Climate Change, 5*(4), 305–314.

Anderies, J. M., Janssen, M. A., & Ostrom, E. (2004). A framework to analyze the robustness of social-ecological systems from an institutional perspective. *Ecology and Society, 9*(1), 18.

Bhattacharyya, S. C. (2013). Financing energy access and off-grid electrification: a review of status, options and challenges. *Renewable and Sustainable Energy Reviews, 20*, 462–472.

Bijker, W., Hughes, T., & Pinch, T. (1987). *The Social Construction of Technological Systems*. Cambridge, MA: MIT Press.

Boait, P., Tembo, B., & Aras, A. (2019). Socioeconomic and environmental impact of a PV-diesel hybrid mini-grid in a remote island of Lake Victoria, Tanzania. *Energy for Sustainable Development, 52*, 85–94.

Boche, A., Foucher, C., & Villa, L. (2022). Understanding microgrid sustainability: a systemic and comprehensive review. *Energies, 15*, 2906.

Bogere, P., Temmen, K., & Bode, H. (2023). Knowledge transfer concepts for microgrids sustainability—the case of East Africa. In *2023 IEEE Global Engineering Education Conference (EDUCON)*. IEEE Kuwait: Xplore, pp. 1–6.

Bukari, D., Kemausuor, F., Quansah, D. A., & Adaramola, M. S. (2021). Towards accelerating the deployment of decentralised renewable energy mini-grids in Ghana: a review. *Renewable and Sustainable Energy Reviews, 135*, 110387.

Carlsson, L., & Sandström, A. (2008). Network governance of the commons. *International Journal of the Commons, 2*, 33–54.

Cegarra-Navarro, J. G., Garcia-Perez, A., & Moreno-Cegarra, J. L. (2014). Technology knowledge and governance: empowering citizen engagement and participation. *Government Information Quarterly, 31*(4), 660–668.

Chiesi, A. M. (2006). Perspectives of network analysis applied to social sciences. *SIS, Atti della XLIII Riunione Scientifica*, CLEUP, Torino.

Cooper, R., & Foster, M. (1971). Sociotechnical systems. *American Psychologist, 26*(5), 467–474.

Cox, M., Arnold, G., & Tomas, S.V. (2009). *Design Principles are Not Blue Prints, But are They Robust? A Meta-analysis of 112 Studies*. Lincoln Institute of Land Policy. https://www.lincolninst.edu/app/uploads/2024/04/1707_925_Cox20Final.pdf

de-Bruijn, H., Herder, P.M. (2009). System and actor perspectives on sociotechnical systems. *IEEE Transactions on Systems, Man, and Cybernetics—Part A: Systems and Humans, 39*, 981–992. https://doi.org/10.1109/TSMCA.2009.2025452

Derks, M., & Romijn, H. (2019). Sustainable performance challenges of rural microgrids: analysis of incentives and policy framework in Indonesia. *Energy for Sustainable Development, 53*, 57–70. https://doi.org/10.1016/j.esd.2019.08.003

Eklund, M., Khalilpour, K., Voinov, A., & Hossain, M. J. (2023). Understanding the community in community microgrids: a conceptual framework for better decision-making. *Energy Research & Social Science, 104,* 103260. https://doi.org/10.1016/j.erss.2023.103260

Energypedia. (n.d.). https://energypedia.info/wiki/Mini-grid_Operation_Models

Gollwitzer, L. (2014). Community-based micro grids: a common property resource problem. *STEPS Working Paper 68.* Brighton: STEPS Centre.

Hatziargyriou, N. D., Kleftakis, V. A., Papadimitriou, C., & Messinis, G. M. (2016). Microgrids in distribution. In: C. Liu, S. McArthur, & S. Lee (Eds.), *Smart Grid Handbook.* Malden, MA: Wiley.

IEA (International Energy Agency). (2022). *Africa Energy Outlook 2022.* International Energy Agency. https://www.iea.org/reports/africa-energy-outlook-2022

IEA, IRENA, UNSD, World Bank, WHO. (2022). *Tracking SDG7: The Energy Progress Report 2022.* Washington, DC. www.worldbank.org

Jack, K., & Smith, G. (2020). Charging ahead: prepaid metering, electricity use, and utility revenue. *American Economic Journal: Applied Economics, 12*(2), 134–168.

Kavanagh, C. (2019). New tech, new threats, and new governance challenges: an opportunity to craft smarter responses? *Carnegie Endowment for International Peace,* Massachusetts.

Kirubi, C. G. (2009). *Expanding Access to Off-grid Rural Electrification in Africa: An Analysis of Community-based Micro-grids in Kenya.* Doctoral thesis, University of California, Berkeley.

Lightfoot, G., & Wisniewski, T. P. (2014). Information asymmetry and power in a surveillance society. *Information and Organization, 24*(4), 214–235.

Liu, G., & Starke, M. (2025). Networked microgrid energy management considering ownership and control structures: a comparison. *Energies, 18*(5), 1115.

Lozny, L. L. (2023). Polycentric governance as a practical strategy for balanced policing: a cross-cultural analysis. *Social Studies,* 22(2).

Motjoadi, V., Bokoro, P. N., Onibonoje, M. O. (2020). A review of microgrid-based approach to rural electrification in South Africa: architecture and policy framework. *Energies, 13,* 2193.

Mottram, H. (2022). Injustices in rural electrification: exploring equity concerns in privately owned minigrids in Tanzania. *Energy Research & Social Science, 93,* 102829.

Muchunku, C., Ulsrud, K., Palit, D., & Jonker-Klunne, W. (2018). Diffusion of solar PV in East Africa: what can be learned from private sector delivery models? *Wiley Interdisciplinary Reviews: Energy and Environment, 7*(3), e282.

Namujju, L. D., Acquah-Swanzy, H., Ngoti, I. F. (2023). An IAD framework analysis of minigrid institutions for sustainable rural electrification in East Africa: a comparative study of Uganda and Tanzania. *Energy Policy, 182,* 113742. https://doi.org/10.1016/j.enpol.2023.113742

Namujju, L. D. (2024). Navigating emergent effects in off-grid systems: Ostrom's design principles and rural energy policy implications. *Energy Research & Social Science, 118,* 103786.

Nathan, H. S. K., Das, S. D., & PS, A.P. (2022). Rural microgrids—»tragedy of commons« or »community collective action«. *Journal of Environmental Policy & Planning, 24*(4), 391–406.

Ngoti, I. F. (2024a). Institutional arrangements and sustainable maintenance management of community-based mini-grids in Tanzania. *Energy Research & Social Science, 115*, 103632. https://doi.org/10.1016/j.erss.2024.103632

Ngoti, I. F. (2024b). The role of sense of ownership in rural community mini-grid management: qualitative case study from Tanzania. *Energy, Sustainability and Society, 14*(1), 1–16. https://doi.org/10.1186/s13705-024-00496-7

Nwanevu, C., Oladipo, S. S., & Ajayi, A. O. (2024). Designing effective policy frameworks for the implementation of microgrids in developing countries: opportunities, challenges and pathways to sustainable energy access. *World Journal of Advanced Research and Reviews, 24*(01), 1279–1294.

North, D. C. (1990). *Institutions, Institutional Change and Economic Performance.* Cambridge: Cambridge University Press.

Odarno, L., Sawe, E., Swai, M., Katyega, M. J., & Lee, A. (2017). *Accelerating Mini-grid Deployment in Sub-Saharan Africa: Lessons from Tanzania.* World Resources Institute.

Ondraczek, J. (2013). The sun rises in the east (of Africa): a comparison of the development and status of solar energy markets in Kenya and Tanzania. *Energy Policy, 56*, 407–417. https://doi.org/10.1016/j.enpol.2013.01.007

Ostrom, E. (1990). *Governing the Commons: The Evolution of Institutions for Collective Action.* Cambridge: Cambridge University Press.

Ostrom, E. (2002). Common-pool resources and institutions: toward a revised theory. In B. Gardner and G. Rausser (Eds), *Handbook of Agricultural Economics*, Vol. 2, Part A. Elsevier, pp. 1315–1339.

Ottenburger, S. S., Cox, R., Chowdhury, B. H., Trybushnyi, D., Al Omar, E., Kaloti, S. A., Ufer, U., Poganietz, W.-R., Liu, W., Deines, E., Müller, T. O., Möhrle, S., & Raskob, W. (2024). Sustainable urban transformations based on integrated microgrid designs. *Nature Sustainability*, 7, 1067–1079.

Pereira, Â. G., & Quintana, S. C. (2002). From technocratic to participatory decision support systems: responding to the new governance initiatives. *Journal of Geographic Information and Decision Analysis, 6*(2), 95–107.

Pérez-Arriaga, I. J. (2022). *Sustainable and Scalable Business Models for Minigrids.* ASR Working Paper. African School of Regulation. https://africanschoolregulation.org/wp-content/uploads/2023/04/ASR-Working-Paper.pdf

Pierce, J. L., Kostova, T., & Dirks, K. T. (2001). Toward a theory of psychological ownership in organizations. *Academy of Management Review, 26*(2), 298–310.

Pitt, J., Diaconescu, A., & Ober, J. (2018). Knowledge management for democratic governance of socio-technical systems. In *The Future of Digital Democracy: An Interdisciplinary Approach.* Cham: Springer, pp. 38–61.

Righi, A. W., & Saurin, T. A. (2015). Complex socio-technical systems: characterization and management guidelines. *Applied Ergonomics, 50*, 19–30. https://doi.org/10.1016/j.apergo.2015.02.003

Roy, G., Ostrom, E., & Walker, J. M. (1990). The nature of common-pool resource problems. *Rationality and Society, 2*(3), 335–358.

Slough, T., Rubenson, D., Levy, R., Alpizar-Rodriguez, F., Del Bernedo Carpio, M., Buntaine, M. T., Christensen, D., Cooperman, A., Eisenbarth, S., Ferraro, P. J., & Zhang, B. (2021). Adoption of community monitoring improves common pool resource management across contexts. *Proceedings of the National Academy of Sciences, 118*(29).

Sovacool, B. K., & Dworkin, M. H. (2015). Energy justice: conceptual insights and practical applications. *Applied Energy, 142*, 435–444.

Sulaeman, I., Simatupang, D. P., Noya, B. K., Suryani, A., Moonen, N., Popovic, J., & Leferink, F. (2021). Remote microgrids for energy access in Indonesia—part I: scaling and sustainability challenges and a technology outlook. *Energies, 14*, 6643.

Tenenbaum, B., Greacen, C., Siyambalapitiya, T., & Knuckles, J. (2014). *From the Bottom Up: How Small Power Producers and Mini-grids can Deliver Electrification and Renewable Energy in Africa*. The World Bank. https://doi.org/10.1596/978-1-4648-0093-1

Trotter, P. A., McManus, M. C., & Maconachie, R. (2017). Electricity planning and implementation in sub-Saharan Africa: a systematic review. *Renewable and Sustainable Energy Reviews, 74*, 1189–1209.

Walker, G., Devine-Wright, P., Hunter, S., High, H., & Evans, B. (2010). Trust and community: exploring the meanings, contexts and dynamics of community renewable energy. *Energy Policy, 38*(6), 2655–2663.

Wentz, J., & Pappalardo, C. (2016). Scaling up local solutions: creating an enabling legal environment for the deployment of community-based renewable microgrids. In J. Jaria-Manzano, N. Chalifour, & L. J. Kotzé (Eds), *Energy, Governance and Sustainability*. Cheltenham: Edward Elgar Publishing.

4.3 Stakeholder Conflicts in Energy Communities

4.3.1 Introduction
Teddy Mangeni

One of the most effective but seldom considered ways of reaching the unelectrified with reliable, affordable, clean, and abundant electricity faster and conveniently is the deployment of microgrids (AMDA, 2022, p. 3). While they play a crucial role in accomplishing the goal of the UN Sustainable Energy for All (SE4All) initiative to provide universal access to electricity by 2030, the microgrid sector faces challenges in realizing that potential. To date, few examples of sustainably working microgrid programmes exist (Peters et al., 2019). All too often, rural electrification and renewable energy initiatives are proposed with grand pronouncements from both donors and government officials. But sadly, it is not uncommon for them to achieve far less than originally envisioned because of a lack of attention to the practical aspects of implementation (Tenenbaum et al., 2014, p. 16). The Africa Research and Teaching Platform for Development (A:RT-D)—Sustainable(A:RT-D)—Sustainable Modular Grids for Grid Stability, an international collaborative research project, recognizes that the introduction of smart grid technology can cause unintended negative effects and that uninformed decisions by the parties involved contribute to conflicts within energy project implementations. Additionally, it acknowledges that technology-intensive engineering projects require similarly intense social interventions to ensure truly sustainable microgrid implementation (Kakande et al., 2019, pp. 6–11). In this chapter stakeholder conflicts in light of the ensuing discussions, field experiences and data are considered with a view to the interdisciplinary scope of renewable energy projects and sustainable renewable energy. In 4.3.2, the project interdisciplinary team discusses the roll-out of the Traffic Light System (TLS)-smart meters in the Tanzanian microgrid community, highlighting the participatory approach, emerging foreseen and unforeseen circumstances, conflict identification and resolution process. In 4.3.3, Acquah-Swanzy examines common-pool resource (CPR) problems, highlighting the related key interventions, and how they impact consumption patterns, resource efficiency, and long-term sustainability of the microgrid. In 4.3.4, using Galtung's ABC triangle model and the Tanzanian community managed microgrid data, Ngoti explores the relationship between ownership and conflict in microgrids and in 4.3.5, in the final subsection, the relationship between participatory conflict analysis and sustainable renewable energy development is explored by Mangeni in light of the varying dimensions of the response, stakeholder participation and energy acceptance.

4.3.2 Interdisciplinary Conflict Perspectives: »Traffic Light System Conflicts«

Teddy Mangeni, Henry Asiimwe, Paul Bogere, Ibrahim Mwammenywa, Josephine Nakato Kakande

Peters et al. (2019) detail the various challenges faced by community-operated microgrids. They state that for microgrids operated by a community, several additional challenges emerge. Finding skillful staff in the community to operate, manage, and maintain the microgrid is often a challenge. Moreover, tariffs are often set way below a level that ensures financial and technical sustainability, and fees are not rigorously collected. The absence of an external operator makes it more difficult for a community-based microgrid to pursue a quasi-commercial approach which at least yields enough revenues to pay for maintenance and replacement investments. The ability to pay is low for the same reasons as given above for customers of commercially operated microgrids. In addition, users perceive the grid as community property and are not always convinced of the need to pay prices that would be required to cover costs. For example, if microgrids run on renewable energies (solar, hydro, wind) so that no operational cost for fuels exists, the population may think the service should be provided for free. Moreover, in the typical design of community-operated microgrids, investment costs do not have to be recovered from users due to external grants, and the community is aware of this. The need to accumulate funds for maintenance and repair is often not fully understood or it is under-appreciated. In addition, incentives and obstacles to enforce payment rigorously are different for community members than for external staff members of a commercial operator; social entanglements complicate rigorous enforcement. Since the management team members themselves are inhabitants of the community, it is difficult for them—due to social relations among the villagers—to apply sanctions against neighbours and friends, especially if those people are encountering financial problems. More so, once one customer starts delaying payment or even defaults without any consequence, bad payment practices spread rapidly among customers (Peters et al., 2019, p. 5).

The observation of the community-operated microgrids discussed did not greatly differ from the challenges faced by the Tanzanian community-managed case presented in Chapter 2 of this book. The structured questionnaires, interviews, focus group discussions, technical equipment measurements, and observations conducted in this remote Tanzanian microgrid village gave insight into the context of the community and microgrid operations. The community-managed microgrid is an eight-year solar-powered electricity project that has been in operation since 2015. The 15 kWp solar PV microgrid project that serves 60 electricity connections (including households, home businesses, a local church, and a primary school) was established as a result of a partnership between an Australian solar developer and the Tanzanian government. Respondents shared that it is one of the three microgrid projects that are still operating out of the 14 installed projects. The project initiation was commenced at the national and international levels. It was later introduced to the village community leaders and residents through the Ministry of Energy and the district officials. A book compiled by the Ministry of Energy served to educate the community about the various stakeholders relevant to the project operation, their enrolment, roles, and the procedures of the microgrid operation.

Discussions with the electricity committee members and the community leaders revealed that despite the 60 electricity user connections (and even more, owing to illegal extensions), the microgrid struggled to meet its maintenance requirements. The batteries were failing and the electricity project did not have enough money to replace them despite having over five years' worth of utility bills collections. This was attributed to the instances of many customers defaulting on the payment of monthly electricity utility fees. It was stated that some community members believed that this project was introduced to the village by their political leader as a gift for their political support, while others believed that it was a donor project. They therefore consumed electricity and refused to pay at the end of the month when the electricity committee members went to physically collect the electricity utility dues, as stipulated in the guidelines book. The electricity committee members were met with verbal, physical, and spiritual harassment at the premises of some defaulters who would insult them, wield machetes, and even threaten curses and witchcraft upon them. Several community residents (former and existing consumers) shared that they were faced with what they felt were unfair situations. They all paid a flat fee for electricity utility consumption regardless of whether they utilized the electricity for household, institution, or business purposes. Others stated that they would pay for their electricity consumption but some electricity committee members would not keep proper records and would later accuse them of non-payment. Due to these conflicts, some users chose to withdraw from the electricity project, while others were cut off from the electricity service as a result of non-compliance. The electricity technician who was charged with disconnecting defaulting users from the microgrid was met with much animosity and resentment. The presenting community context translated into uninformed consumption patterns, the project's inability to meet maintenance demands of the microgrid, as well as failing and poorly maintained technical equipment. This, in turn, was manifested through insufficient, fluctuating access to electricity services and stakeholder conflicts. In the initial phases of data collection, a sense of hopelessness was expressed towards the future of the microgrid project, such as in the following:

> This village is ok, it is standing but also the batteries are faulty, the batteries expired, but they are still working, but, even now there is no electricity, right now. This night because their batteries died, they don't have much power. When the battery die, there is no capacity to store charge. It cannot store electricity, and within a few years, this

> may also go down completely. There is no any plan, but they decide to find someone to help us, a donor, to help us with this project because we do not have money to buy batteries.
>
> *Interview with electrician*

It was repeatedly said that a self-payment metering system that would enable individual electricity users to track their consumption levels would go a long way in addressing the bulk of the challenges that the communities faced. The community respondents reasoned that this would address the sense of inequity that comes from everyone paying a flat rate without knowing how much electricity one has consumed and would minimize the several emerging conflicts and non-payment concerns.

> (...) we need electricity full time. The second need is for the grid to have prepaid meters, this will remove stories of conflict and pangas... because if one has a meter, they use their electricity paid for and if they are not able to, it goes off. Will they look for the committee member when their electricity goes off, where will they find the committee member to fight with?
>
> *FGD with electricity committee members*

In light of the challenges and desired solutions expressed by the microgrid community, an interdisciplinary team comprised of experts in engineering, didactics of technology, and social sciences, community engagements, the introduction, and implementation process of a »smart metering—Traffic Light System (TLS)« came forth as a solution to the challenges presented by the community. The TLS invention, led by the A:RT-D Grids team, offered a solution to address, minimize, and even eradicate many of the challenges that this microgrid community was facing.

The TLS metering system presented the community with the ability to alleviate the circumstances of the concerns raised; a project interdisciplinary team comprised of engineering, didactics of technology, and social sciences experts participated in the roll out of the metering system in the microgrid community. The designed TLS provided a low-cost consumer module for demand-side load monitoring and management in microgrids. It is a system that monitors the consumer's load and controls the load to maintain the microgrid stability and to avoid blackouts during times of low power generation. This is done through clear, colour-coded visual cues that advise and guide residents towards more efficient energy choices. The two-way communication helps the utility to both monitor the demand as well as to manage the consumption of electric power based on power generation (Mwammenywa et al., 2021, p. 2; Mwammenywa et al., 2022).

4.3.2.1 Technical Dimensions of the TLS Interdisciplinary Intervention

The TLS system contains the customer modules (the smart meters with traffic lights) and the host server which is connected to the solar microgrid power plant. The customer module has three lights, which are controlled by signals sent from the host computer using wireless LoRa technology. The LoRa transceivers enable bidirectional communication between the host computer and the customer modules. The customer module meters the load of a particular user and compares it with a control signal received from the host computer. If the load is greater than the limit, a red light comes on, indicating a user should reduce the power consumption or risks a power disconnection from the grid. If load is around 20 % from the limit, the yellow light comes on, indicating to a user not to add any load. The green light comes on only when the load is much lower than the limit.

The host computer calibrates the limit signal based on the amount stored in the batteries and the power generated by the solar PVs. This calibration is based on the Fuzzy Logic algorithm of artificial intelligence. The host computer also records the power consumption of each connected user, providing the data which can be used to calculate electricity bills with respect of usage instead of the flat monthly bills. Fig. 50 below shows a smart meter TLS architecture.

The smart meters were installed to the connected electricity users. The community members were trained about the meaning of the traffic-light signals prior to the installation of the meters. During the training, special emphasis was placed on the importance of regulating the electricity demand based on traffic-light colour so as to improve the reliability and sustainability of the microgrid. Moreover, the implication of automatic disconnection from the grid for those not adhering to the traffic-light rules were discussed.

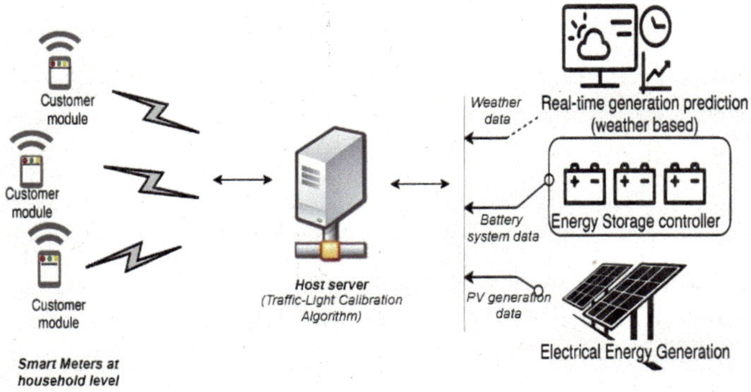

Fig. 50: TLS smart metering architecture (Namujju et al., 2024, p. 2).

Furthermore, an analysis of the load distribution of the three distribution phases from the plant to the connected users, using a Mavowatt power quality analyser, revealed that one of the phases was overloaded when compared to the other two phases. This phase also had a lower reliability, incurring more outages than the other two over an approximately two-week measurement period. Poor quality and unreliable power supply leads to customer dissatisfaction and customer attrition and can damage system and customer equipment. It could also be a cause of conflict for the microgrid technical personnel and/or operators who would be put to task to explain why some customers have power while others do not. Thus, it was important to ensure that periodic line balancing is done to avoid overloaded lines and equipment, which shortens the lifetime of the equipment (Kakande et al., 2024). This adversely affects microgrid sustainability from both a reduced revenue perspective, as a result of unmet demand and possible customer loss, as well as from the increase in operational and maintenance costs due to more frequent equipment repair and replacement requirements for the overloaded phase(s). Hence, the microgrid technical personnel were guided on handling line balancing.

4.3.2.2 Social Sciences and Didactics for Technology Dimensions

The introduction of the microgrid in this community stemmed from a top-to-bottom approach, i.e., from the international, national, and district levels to the community local leaders and eventually the community users. The community was not engaged in consultative meetings and assessments, and therefore the developers did not understand the context, culture, community beliefs, gender notions, roles, and related schedules of the community among other aspects that were relevant to the project. Hence the ensuing emerging ambiguities and conflicts that were reported.

Citizen participation is a categorical term for citizen power. It is the redistribution of power that enables the have-not citizens, presently excluded from the political and economic processes, to be deliberately included in the future. It is the strategy by which the have-nots join in determining how information is shared and which goals and policies are set. In short, it is the means by which citizens can induce significant social reform which enables them to share in the benefits of the affluent society (Arnstein, 2019, p. 1).

Interviews and focus group discussions conducted with community respondents guided the identification of the stakeholders in the microgrid process and their roles. These included the following:

- Village Executive Officer (VEO)—An administrative politically appointed leader in charge of the overall coordination of the village and oversight of the microgrid operations.
- Local council community leaders—Community elected leaders. They oversee the coordination of the village operation in liaison with the VEO. Some of the members are part of the financial signatory committee of the microgrid community. Together with the VEO, they oversee and guide the election and operations of the electricity committee.
- Electricity committee—A team comprising eight elected members who conduct the day-to-day operations of the microgrid. These activities include the physical collection of monthly utility bills, the deposit of collected monies in the microgrid community bank account, complaints, feedback, and compliance management, among others.

- Electricity technician—A person hired and left in place by the microgrid developer and the Ministry of Energy officials. This person is in charge of the technical maintenance of the community-managed microgrid.
- Electricity users—Community members that are utilizing the microgrid electricity. They are obliged to pay for the utility bills. They vote for the electricity committee and the leaders. They are expected to hold the electricity committee and the community leaders accountable for the microgrid operation decisions undertaken and participate in project evaluation and maintenance.

The identification of the community stakeholder-role matrix guided the involvement of the community in the introduction, enrolment, and implementation of the TLS. It guided the compilation of the community engagement strategy for the TLS project. The administrative, local leaders, and electricity committee members were taken through education and discussion sessions regarding the TLS system. They then mobilized the electricity users to participate in the project activities such as education sessions and acceptance of the system installation in the premises. The hierarchical entrance into the community contributed to the legality and acceptance of the TLS project. The leaders moved with the technicians to each household during the installation of the LoRa meters. They further educated the community members about the meters and responded to any inquiries and concerns that arose.

Despite the technical challenges that arose after the installation of the meters, the communities did not turn hostile, as seen in this field citation:

> No, the people complained, but nobody touched the gadgets. They associate them with the government and therefore cannot risk touching or removing them. They assume everything has been put with the approval of the government, so they are scared of tampering with them. (…) we called for a meeting and the electrician talked to us and explained to us. We all understood. All the people entered the container and we were explained to, how the meters work and also the

> different challenges of the container. And we all seem to understand. The remaining problem is that we were explained to that our electricity has three electricity loads and one of them is heavily loaded. We were guided that what we need to is to balance these loads so that we can stop these electricity blackouts. (FGD with community leaders and electricity committee).

The microgrid community looked to their leaders and through liaison with the technicians and the A:RT-D sensorics technical team, participatory solutions were born. An interdisciplinary and participatory approach to the technical process ensures that all facets relevant to the success of the project are identified and employed. It contributes to meaningful stakeholder participation that empowers them with the capacity to sustain and maintain a positive outcome of the microgrid process.

4.3.3 The Impact of Nudging and Centralized Control on Sustainability in Flat-Rate Microgrids—A Common-Pool Resource Perspective

Henrietta Acquah-Swanzy

4.3.3.1 Introduction

Most community microgrids are isolated and not connected to other grids. This means that the available capacity is to be shared among the community members and there is no alternative supply, not even at a high cost, to top up during peak seasons. The reliance on natural resources like solar and hydro also means that there is uncertainty in the system capacity supply at each point in time. It has been described as a common-pool resource (CPR).

Understood as CPRs, microgrids are marked by characteristics such as rivalry whereby the generated power is finite and can be depleted through battery storage, consumption or use. Once users are connected, it is difficult to exclude them from using power generated without a technical or in some cases social intervention. Exhaustibility of the microgrid means that, if not managed sustainably, the system can degrade and shut down.

These features of the microgrid can lead to a »tragedy of the commons«, whereby individuals act in their self-interest without considering the impact on the overall microgrid, leading to potential overuse and depletion over time. There are also potential challenges that might result from individuals and groups in the community not being able to coordinate in order to achieve the sustainable use of the microgrid.

The greater the uncertainty in the threshold of a CPR, the higher the likelihood of consumers not extracting at sustainable levels (Kidwai & de Oliveira, 2020; Maas et al., 2017). Some suggested solutions to improving sustainability chances include different forms of communication and coordination (Gehrig et al., 2019; Travers et al., 2011) and introducing forms of monitoring and punishments (Maas et al., 2017; Rapoport & Au, 2001) with added layers of willingness and enforceability.

This study examines a community microgrid as a CPR problem, where unrestricted access and shared usage create risks of overconsumption and inefficiency. By leveraging behavioural insights, we investigate the role of two key interventions: centralized control with enforcement measures and behavioural nudges, the latter of which aim at influencing community energy consumption. To this end, we examine key findings from two publications, which applied three different methodologies. We assess how these interventions impact consumption patterns, resource efficiency, and long-term sustainability of the microgrid.

Central control in a microgrid context typically involves an external authority, like a utility or microgrid management, setting strict consumption rules or penalties to manage energy use. This often relies on enforcement mechanisms (e.g., penalties for overuse) to maintain compliance as well as technological interventions which can lead to more predictable outcomes in resource usage and efficiency.

Nudging, which involves the use of behavioural prompts that guide decision-making without force, on the other hand, encourages community members to follow shared expectations voluntarily. Nudging relies on the ability of consumers to understand and correctly apply the nudges and the willingness to act accordingly.

We ask the question: In a CPR framed microgrid, what are the effects of centrally controlled distribution measures and nudging on consumption, efficiency, and sustainable foresight?

4.3.3.2 Methodology

To assess and test the effectiveness of nudging and central control on the efficiency and sustainability of microgrids, researchers employ a variety of approaches. In this section, we present three key methodologies: theoretical analysis, laboratory experiments, and field experiments. Each approach has its strengths and limitations, which are explored below to provide a comprehensive understanding of their applicability to microgrid research.

Theoretical analysis involves the development of mathematical models and simulations to predict the behaviour of energy consumers and the microgrid's performance under different intervention scenarios (e.g., nudging or central control). These models are based on principles from game theory and behavioural as well as classical economics. Classical game theory models allow us to analyse how individuals make decisions about energy consumption in shared systems, accounting for their self-interest versus collective outcomes. Behavioural game theory models simulate the performance of microgrids under various demand-side management (DSM) strategies, such as behavioural nudges and central controls, where the »behavioural« part emphasizes psychological components on how consumers respond to feedback, social norms or other signals. Theoretical models allow researchers to derive predictions for multiple scenarios, thus providing a cost-effective way to forecast the potential outcomes of interventions before implementation, saving time and resources. That said, such models often rely on idealized assumptions such as, for example, rational consumer behaviour and symmetry, which may not align with real-world complexity.

Laboratory experiments create controlled environments where participants assume the role of an energy consumer in a microgrid. These experiments test behavioural responses to nudging or central control under tightly regulated conditions, allowing for the collection of precise, high-quality data. Laboratory experiments allow the direct observation of how participants react to different interventions, providing valuable insights into decision-making processes with the added advantage of replicability, ensuring the reliability of results. Yet, these experiments might be limited in terms of their real-world applicability (*external validity*) and scale.

Field experiments implement interventions in real-world microgrids and observe their impacts on actual consumer behaviour and system performance. Here, interventions have the advantage of capturing actual consumer behaviour and system dynamics, providing the most accurate representation of intervention impacts. They can also account for external factors, such as cultural norms, social dynamics, and infrastructure limitations, which influence the effectiveness of interventions. However, they come with some ethical concerns as well as complexity and cost as they require substantial resources and infrastructure.

4.3.3.3 Results

In this section we refer to results from two articles. In the first, Acquah-Swanzy & Hehenkamp (forthcoming) derive theoretical predictions and use lab experiment results to examine the effect of nudging and redistributional control in microgrid setups as a simultaneously repeated CPR game. In the second, Namujju et al. (2024) present evidence of the effect of a combination of nudging and central control on the peak period consumption and general consumption patterns. In both studies, visual cues in the form of traffic-light signals are used.

Theoretical results. In a microgrid flat rate pricing framework, Acquah-Swanzy & Hehenkamp (forthcoming) use a CPR model based on Budescu et al. (1995) as a simultaneously repeated game. Assuming symmetric agents, the authors show that, in the presence of risk in the resource supply, there are two main consumption points: either consumers take the positive minimum and consume an equal share or, in a riskier setting, consumers consume above the equal share. Nudging results are signal based, whereby positive and negative signals will at least increase or maintain the consumers equal share amount without making them worse off. A central control redistribution setup also showed that, above a certain threshold of punishment, consumers will consume the equal share since they will be worse off otherwise. The central control, however, results in better efficiency and sustainability as it is more predetermined compared to the nudging.

Laboratory experiment results. A laboratory experiment was subsequently conducted based on the theoretical setup provided by Acquah-Swanzy & Hehenkamp (forthcoming). It turned out that the nudge treatment has a significant impact when it comes to directing consumers to the right mode of consumption. A positive signal results in higher consumption, and a negative signal leads to a reduction in consumption. While the nudge resulted in better resource efficiency than in an uncontrolled situation without nudging, sustainability was not significantly improved. While having significant effects on resource efficiency and sustainability, the central control treatment did not lead to better social efficiency when compared to the nudge case or the baseline treatment with neither central control nor a nudge.

Field experiment results. Namujju et al. (2024) conducted a field experiment that combined nudging and central control within an innovative energy DSM system. Their study setup also mimicked a CPR with an isolated microgrid and flat rate pricing. They provided nudging cues, and upon ignoring negative nudge signals, consumers' supply was forcibly reduced. They found this to entail a significant reduction in peak period consumption and a change in power consumption patterns.

4.3.3.4 Conclusion

We have presented three research methodologies to study the effect of nudging and central control on microgrid consumption and sustainability, analysing this as a CPR system. The results showed a positive effect of nudging in directing consumption patterns under all three methodologies. Central controls were found to improve resource efficiency and sustainability outcomes in both the laboratory and the field experiments.

4.3.4 Ownership and Conflict in Community-Based Microgrids: Insights from Rural Tanzania

Irene F. Ngoti

The sustainability of community microgrids in rural areas of developing countries is a growing concern for energy stakeholders. Despite having advantages such as local participation and empowerment, community-managed microgrids also face technical, governmental, political, financial, and social-technical challenges (Babayomi & Okharedia, 2019; Odarno et al., 2017; Ogeya & Lambe, 2025; Zigah et al., 2023). Since community-managed microgrids are also exposed to conflict (Asiimwe et al., 2025), it is vital to explore how ownership can be a centre of conflict and hence threaten microgrid sustainability. Understanding how conflict among microgrid users is influenced by microgrid ownership is one way to contribute to the design of sustainable community-based microgrids.

Due to financial constraints among rural communities in developing economies, including Tanzania, most microgrid projects are externally financed from the design to the implementation phase. Beneficiary communities participate in such projects through sweat equity and by offering land where such projects are installed. The ownership structure is made known to all stakeholders from the beginning, together with operational, maintenance, and management matters, to avoid confusion in the latter stages. In case of transfer of ownership from the developer to a third party (e.g., community, government, or utility company), regulatory frameworks and the developer often have a say on it (Mini-Grids Partnership, 2020). Ownership structure and status, therefore, have an important role in the sustainable governance and management of rural community microgrids.

This study explores the relationship between ownership and conflict in communally managed microgrids in rural Tanzania. It employs Galtung's ABC triangle model (Galtung, 1996), whose elements—attitudes, behaviours, and contradictions—are utilized to evaluate how microgrid ownership structure can result in conflict.

4.3.4.1 Study Framework

4.3.4.1.1 The ABC Framework and Conflict

This study is anchored in the ABC framework (Galtung, 1996) as its conceptual foundation for understanding conflicts in community microgrids and how ownership in community-managed microgrids initiates conflict. »Conflict« originates from the Latin word »conflictus«, meaning collision or clash. There is no single definition of »conflict«, as many people attempt to define it based on their respective subject of study and various characteristics (Lyamouri-Bajja et al., 2011). Fisher et al. (2000), for instance, define conflict as the relationship between two or more parties (individuals or groups) who have, or think they have, incompatible goals, while Galtung (1996) considers conflict as a dynamic process, in which structure, attitudes, and behaviours are constantly changing and influencing one another. The latter definition is particularly valuable in this study for unlocking fundamental elements of conflict. Based on this definition, the driving factors behind conflicts are represented by the ABC triangle. This model identifies attitudes (A), behaviour (B), and contradiction (C) as three basic components of conflict, which are described in Fig. 51.

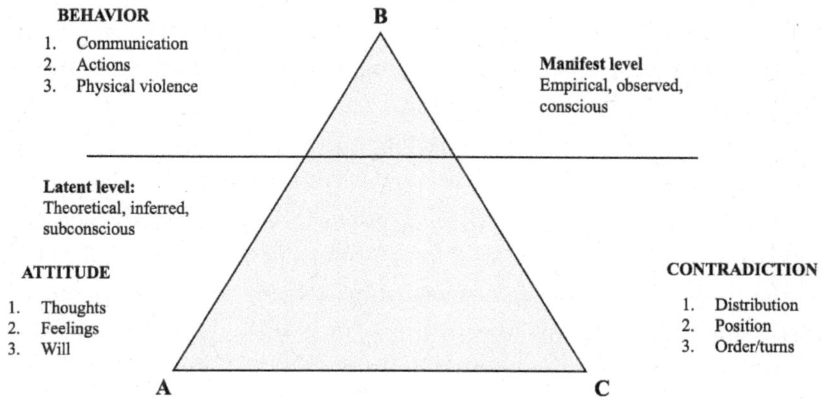

Fig. 51: The conflict triangle (Galtung, 1996, p. 72).

Attitudes are linked to thinking, perceptions, feelings, and beliefs that actors have within themselves, against each other, or towards a third party. This aspect of conflict is often hidden, shaped by personal perspectives, and tends to differ across the parties involved (Lyamouri-Bajja et al., 2011). Behaviour, the most visible aspect in the model, refers to actions by the conflicting parties. This can be verbal, physical, or both. The contradiction aspect relates to the matter(s) over which disagreement has emerged. It describes the actual incompatibility of goals among parties and considered as the root of conflict. The issues that the model refers to under the contradiction aspect are distribution, position, and order. The three aspects are interrelated, and any of them can be the origin of conflict (Lyamouri-Bajja et al., 2011).

4.3.4.1.2 *Applying the ABC Model to Ownership and Conflict in Community-Based Microgrids*

The ABC triangle provides a comprehensive framework for analysing conflict. Once applied to a microgrid, this model explicates how ownership can initiate conflict among microgrid users, which may pose a threat to both microgrid governance and sustainability. It provides a way in which microgrid ownership moulds the attitudes and behaviours of electricity users, which leads to conflict within rural energy communities. Applied to a microgrid, the model enables a mapping of how ownership structure impacts microgrid users' attitudes towards microgrid governance, how these attitudes manifest in improper behaviour, and how contextual factors magnify different situations.

(A)ttitude: Microgrid Ownership Legitimacy. Ownership in a community microgrid is not just a legal form of ownership; it can also be seen in the presence of a »sense of ownership« (Ngoti, 2024). The nature of ownership shapes the way users consider their roles, responsibilities, and contributions towards the microgrid project. Users' attitudes towards microgrid ownership, being good or bad, are conceived from the inception of projects. For example, if a microgrid is considered community-based, but maintenance management relies entirely on an external company with no financial contributions from microgrid users in the community, then mem-

bers may raise questions about their legitimacy in microgrid ownership. Alternatively, if users have no legal ownership but are responsible for setting tariffs and maintenance management, their perception towards ownership may be positive and vice versa. Attitude in the ABC model, therefore, depicts how microgrid users view or interpret the ownership concept, which can raise conflict or not.

(B)ehaviour: Ownership and Day-to-day Participation. Ownership structure has an impact on day-to-day activities pertaining to microgrid management, such as tariff payment, reporting faults and illegal connections, attending meetings, and usage of allowed electricity appliances. Microgrid users are therefore likely to behave positively by complying with said day-to-day activities when microgrid ownership is perceived as meaningful or legitimate; otherwise, inappropriate behaviour will surface and cause conflict. This aspect of the ABC model is often visible as microgrid users can no longer hide their attitudes towards ownership-related matters and thus decide to unmask their attitudes through practising inappropriate behaviours.

(C)ontradiction: Ownership Standards and Microgrid Governance Realities. Community microgrid ownership identifies beneficiary communities to oversee microgrid operations and management collectively (Maier, 2007; Palit & Chaurey, 2011). This can be practically different and thus create a contradiction, as what seems to be presented theoretically is not actually implemented or is only partially implemented in practice. For example, the domination of external parties on the microgrid operational and management decisions may create a contradiction between what community ownership is and governance practices. Such contradictions may remain dormant until alarming events such as equipment failure bring them to the surface.

4.3.4.1.3 *Ownership-Conflict Reconciliation in Community Microgrids*

The ABC model provides a lens for understanding how the structure of microgrid ownership can be a source of conflict among microgrid users in rural community settings. Microgrid ownership is not only about legal possession of the microgrid but also about the perceptions and relationship of microgrid users towards the project (Ngoti, 2024). The model unfolds how microgrid users' attitudes influence their behaviours on different issues and how contradictions in microgrid governance can raise conflicts, which in turn undermine rural community microgrid projects.

4.3.4.2 Methodological Approach

This study is based on field visits to a Tanzanian microgrid during May and June 2022. A qualitative approach was employed where 10 semi-structured interviews were conducted on the following interviewees: 3 Village Electricity Committee (VEC) members, 1 village council member, 1 technician, and 5 normal electricity users. This microgrid was selected based on its ownership structure and because it was the only surviving community-managed microgrid in Tanzania. The interviews focused on the users' insights on the microgrid maintenance management, institutional arrangements, challenges, and successes towards sustainability. The interviews were later transcribed in their original language, Kiswahili, and later coded using MAXQDA 2022. The interviews were thematically coded based on how interviewees described ownership, conflict, and governance during data collection. These themes were therefore data-driven. The transcripts were later translated into English for analysis purposes.

4.3.4.3 Results and Discussion

4.3.4.3.1 Attitudes Towards the Microgrid Ownership Legitimacy

Interviewees expressed their views on how they felt about the status of microgrid ownership. They were all aware that they had no legal entitlement to microgrid ownership, but most users had a high sense of ownership towards this microgrid project (Ngoti, 2024). Microgrid users were excluded from having financial obligations to the microgrid's operations and maintenance (e.g., paying a technician's salary and maintenance) in the first three years when a third-party company—the electrical company—was responsible for such activities. They were told that the electricity was to be consumed freely, and there was no need to pay for it.

> ...Most users have the same thing in mind, and you will hear them saying that aah, when they brought the electricity for us, it was for us to consume it freely...
>
> *(Member of the VEC)*

This idea was instilled in people's minds during the design and implementation phases. It then changed users' attitudes towards microgrid ownership as they felt excluded from a very important matter of any community project, that is, paying tariffs from the very beginning. The attitudes of the microgrid users were shaped by what they were instilled to believe from the beginning of the project: namely that it was not theirs. Such beliefs became a source of division later when they were to start paying tariffs. The status of ownership in the microgrid was not inclusive, which caused negative attitudes from the users towards it.

4.3.4.3.2 The Microgrid Users' Behaviours Towards Ownership Status

Behaviour, in the ABC model, is the most visible aspect among the three, which can be manifested verbally, physically, or both (Lyamouri-Bajja et al., 2011). In the village, this was reflected in how microgrid users acted in response to their attitudes towards ownership status and surrounding context. The fact that some users stuck with the notion that they were not to pay the tariffs as they were told in the beginning affected their behaviour in most of the implementation phase. Some users' reactions were not positive, both verbally and physically, especially on the issue of tariff payment compliance. Some interviewees had the following to say on this matter:

> ... Sometimes you even encounter swords when collecting tariffs. Someone might say, »Every day it is you who does the collection, is this your father's container! ...«
>
> *(Member of the VEC)*

> ... When this electricity project came, people took it for granted because they did not labour to apply for it. It is something that never

> existed in their dreams, which is why they see it as something that has no value. You see. They despise it…
>
> *(Electricity consumer)*

The two voices clearly show that this nature of conflict through verbal and physical reactions could hardly happen if users were the legal owners of the microgrid, who laboured hard from raising funds to finance it and finally use it.

4.3.4.3.3 Contradictions between Ownership Status and the Community Microgrid Governance

Users in the community were not happy with how microgrid governance was centralized, despite the day-to-day management of the microgrid over four years, since everything concerning management and operations was left to the VEC and the village council after the electrical company left. With a high sense of ownership despite the non-existence of legal ownership (Ngoti, 2024), interviewees were frustrated as to why they could not have total control over the microgrid since there was no external support from the central government and its partners.

> … Some problems were caused by the government, I say by the government because, for example, when these containers were installed, technicians were paid well by the company. There was no problem. But since they left, there have been problems sometimes, to the point of losing trust. The technician says the key is in Dodoma. Now you ask yourself, how is it that the container is in the village, and the key is in Dodoma? It's not true. Community members are therefore forced to pay the return fare to Dodoma but you know in reality no one went to Dodoma…
>
> *(Village council member).*

The microgrid container key was still with the Ministry of Energy and Minerals, and users had to pay a technician to get the key more than 50 miles away every time there was a breakdown. This contradiction created a fuss of noncompliance with tariff payments among users because, as symbolic owners of the microgrid, they felt entitled to having full control over it.

The results also show that users experienced a contradiction that later led to conflict when they thought that things would go smoothly, as they did during operation and maintenance by the electrical company, because it was a project that the whites gave them. When breakdowns that needed maintenance and repair occasionally occurred, conflicts accelerated as they questioned why they would pay for unreliable electricity services.

4.3.4.3.4 *Interactions of ABC Aspects in the Community: From Misalignment to Conflict*

The study explores the relationship between ownership and conflict in community-managed microgrids in Dodoma, Tanzania, through the ABC framework. The three aspects of this framework influence one another and rarely work in isolation. The attitudes of microgrid users supported a collective sense of ownership (Ngoti, 2024), but circumstances created constraints, such as controlling the microgrid container key. The resulting outcomes were behavioural contradictions and conflicts. Again, some users' attitudes in the community supported ownership with an external finance assistance mindset, but the context offered a sense of ownership with financial responsibilities, which produced behavioural contradictions among members and eventually conflicts. Generally, the existence of conflict in the village was not directly due to bad attitudes or behaviour among users but more due to misalignment of one or more of the ABC triangle aspects.

4.3.4.4 Conclusion

Deploying the ABC framework, this study confirms that microgrid ownership is not just about the right to possess a microgrid but is crucial to mould users' attitudes and behaviours, together with controlling any rising contradictions that may result in conflicts. A sustainable microgrid governance requires alignment between users' beliefs about the status of microgrid ownership, which also affects their behaviours towards the entire microgrid life cycle.

4.3.5 Participatory Conflict Analysis and Community Development

Teddy Mangeni

4.3.5.1 Introduction

Regardless of attempted neutrality and impartiality, interventions by development-humanitarian agencies will always affect conflict dynamics and balance of power. Impacts may be either positive or negative and to a greater or lesser extent. There is ample evidence that the social, economic, and political impacts of interventions can amplify grievances, incentivize conflict, or exacerbate tensions (Ware & Laoutides, 2021, p. 675). Participation is a widely accepted principle of development: people should be involved in decisions that affect them because it leads to better decisions, it motivates them, and, ultimately, it is just (Fisher et al., 2000, p. 147). Most development aid systems function top-down, i.e., along externally defined lines, bypassing people's own creativity, capacities, histories, and sense of value. This inevitably has a major impact on the minds and aspirations of people. Instead of people and their values, macroeconomic aggregates and technologies are treated as the core of development. In such circumstances, it should not be surprising when people resist these reductionist schemes—whether by being passive, obnoxious, fundamentalist, cynical, racist, or violent (Uvin, 1996, p. 34). Careful analysis of the context, actors, and dynamics of a conflict is thus required. A careful analysis of the project design then tries to anticipate the interaction between potential interventions and conflict, to adapt the proposed design to minimize harm and maximize positive impact (Ware & Laoutides, 2021, p. 675).

4.3.5.2 Participatory Conflict Analysis and Sustainable Renewable Energy Development

In the quest to contribute to sustainable development through SDG7—ensuring access to affordable energy—renewable energy projects are not excluded from this predicament of double-edged conflict dynamics. Renewable energy sources provide a promising alternative, and their deployment is thought of as an important energy policy priority for many countries. However, such policies must overcome technical, institutional, social, and economic barriers as they manifest localized environmental impacts—extensive land and water use requirements, landscape alteration, change of rural lifestyle, noise, etc. This multiple character of renewables renders the choice between different development proposals a laborious and complex task (Polatidis & Haralambopoulos, 2004, p. 1253). When the rural electrification projects are designed by a central governing body and proceed without consulting the beneficiaries, the project capacity falls below the desired levels (Duran & Sahinyazan, 2021, p. 2). Of course, publicly funded projects require dedication, planning, and a transparent partnership with the implementing organization. Yet, the sustainable management of renewable energy projects is commonly exasperating for the stakeholders involved because the aim of implementation and responsibilities are not sufficiently clarified. Consequently, this makes the projects in many ways uncertain to manage effectively within SSA (Ikejemba et al., 2017, p. 234). Many small and medium-sized PV projects have been implemented; however, despite the potential of these projects, most of them have been left to deteriorate with no maintenance, no cleaning, no repairs, and, more importantly, no sustainable management method that encompasses them all. Some of these projects are precipitously discontinued during planning and sometimes during implementation, leading to the dissipation of limited resources such as funding and time. This situation has created a negative attitude towards the acceptance of sustainable and clean energy usage within the minds of the masses, making it difficult for governments of these developing countries to adopt renewable energy technology (Ikejemba et al., 2017, p. 235). It is crucial for a better understanding of people's responses to renewable energy technology to reflect upon how people make sense of the new, of change, specifically by paying particular attention to the role of conflicting representations at individual, group, inter-group, and societal levels in the acceptance of or resistance to change regarding renewable energy technologies (Batel & Devine-Wright, 2015, p. 9).

4.3.5.3 Conflict Analysis in East African Microgrid Communities

Interviews, focus group discussions, observation conducted in the 14 microgrid communities in Uganda and Tanzania, and document review of stakeholder correspondence and microgrid implementation policy continually brought forth aspects of participation, conflict dynamics, and the resulting effects on community development and sustainability of microgrid projects as set out above.

The microgrid projects, as characteristically intended, provided remote microgrid communities with the opportunity to access electricity. Based on hard-to-reach locations (mountainous, islands, and/or distant), sparse populations, and developing national government economic status of these communities, the respondents—which included regulatory authority staff, microgrid operators, community leaders, and members—shared that the communities did not expect to be connected to the national electricity grid in the near future. The context of these rural communities' circumstances made it prohibitively expensive and technically challenging for the national government to extend electricity to these areas. The data revealed a generally elated and anticipatory response to receiving electricity in the communities. However, despite the excitement, the microgrid electricity projects presented a novel concept of electrification to most community residents. Before microgrid electrification, the communities depended on coal, a few generators, and stand-alone home solar systems for electricity. They stated that these means were expensive and unreliable. The stand-alone solar systems would provide electricity for limited hours of minimal lighting connection and phone charging. During rainy seasons, they would not work at all. Most of the community residents, therefore, doubted the ability of the solar microgrid energy to provide consistent and sufficient electricity to meet their energy needs. The study results show that the reasons explaining the varying dimensions of the response to the uptake of the electricity, levels of community development, and access to energy for all could be traced in the information and stakeholder participation levels in the microgrid processes as discussed in the next section.

4.3.5.4 Contexts of »Power« Acceptance and Community Development

A statement by a community liaison officer in one of the microgrid villages in Uganda captures some of the dynamics and responses that were evident in most of the microgrid villages:

> People are very willing up to now. I think people would start workshops and these maize miller machines. Some people bought it but they are just waiting for connection. And maybe these juice making machines, I am seeing some group organized themselves, then they bought it already, they are just waiting for three phase connection. The women are so involved now, like for these simsim paste millers, I have a group of women also that they report to me, they bought it already but where to put it and the power they have seen it is not enough for this simsim paste. But I convince them that by the end of this month, they will bring the form, I will fill the form with them, then the tree phase connection will be done for those people. I also don't know for sure when this three-phase connection will start.

The field data analysis revealed that the introduction of electricity to the remote villages was met with high expectations. Community development was anticipated. Hospitals, schools, business, and households imagined a world of limitless possibilities that came with this electricity. And yet, the resulting community development and responses to the electricity projects differed within and among different communities (Mangeni, 2024, pp. 168–170).

Euphoric acceptance of the electricity projects was exhibited at the start by most of the microgrid communities. This was characterized by excitement driven by the initial good will of the operators that provided free electricity connections and months of consumption without fees in some communities. Also, respondents stated that there was a general lack of comprehensive information about the electricity and process capacities and limitations at the project's outset. Many customers said that they clamoured to connect to the electricity and bought several electricity gadgets anticipating the limitless ability of the electricity. Only later did they realize the following: that one was required to pay monthly utility bills including consumption and service fees; that the limited voltage capacity of the electricity could only power lighting and two sockets despite the numerous pieces of electrical equipment they had spent money on; that the operator in some communities required them to complete the connection fee at a short-term pressured rate, as opposed to the lengthy leisurely installment period that they had been led to believe and expect; and that the institutions, such as schools, hospitals, district headquarters, in some microgrid communities that had given over land for the construction of power plants did not receive any special consideration, and that they, too, received the three light bulbs and two-socket connection despite the need for a greater voltage connection. In some instances, even when housing the plant, these institutions were not connected to the electricity, citing operators having exceeded connection capacity. The communities started to feel that they were only »cosmetically« involved, if at all, in the development of the electricity projects, and that many aspects, such as equipment installations, amounts and mode of payments, did not speak to the contexts of community livelihood, income generation patterns, and general community beliefs. Hence, the period of euphoria came to an abrupt end; emerging and widening inequalities and sense of unfairness cropped up in its place and were voiced. Animosity among stakeholders, abandonment of the project, and destruction of solar project equipment, including the stoning of solar panels, among other responses, were registered in some villages.

Some community members chose trading-off as a response, as they were not happy about some aspects of the microgrid projects. Some of these stated aspects included: the introduction of additional service fees and connection costs that were not originally declared by operators and/or agreed upon with the electricity users at the onset of the electricity use contracts; and the changing terms of conditions, e.g., changing the ratios of cost-sharing for productive use of energy, to make electricity users pay a higher ratio than they were originally led to believe. While users continued to use the electricity and related services, they were not happy. They therefore boycotted and did not participate in the mobilized and »supposed«

consultation project activities and did not attend meetings. Statements like »*(…) Why would I attend those meetings? Whenever you attend the meetings, those people keep changing their statements!*« were commonplace. Community members stated that when the next alternative project to bring electricity would come along, such as the national electrification drives or larger capacity stand-alone solar batteries, they would abandon the project.

Divergent responses were revealed in other instances where some electricity users desired electricity but could not connect to it due to the limiting operator capacity threshold or could not afford to pay the monthly utility bills. In some communities, technicians and project operator staff roles were not formally documented or supervised. Activities that resulted included illegal electricity connections, refusal to pay monthly utility bills, financial fraudulence and manipulation, and abandonment of duty.

Exclusion as a response to the existing and emerging conflicts in microgrid communities was revealed by the field data in some microgrid communities. This was both voluntary and involuntary. Project implementation conditions were one source of discontent, which included dishonesty and the changing of terms and conditions by microgrid operators, manipulative and selective information sharing, inclusion and participation platforms; some electricity users and their leaders chose to outright reject and not engage in the electricity projects, while others withdrew their participation in the midst of project implementation. In other instances, stakeholders, such as payment defaulters, those with illegal connections, wayward staff, among others, were cut off from participation in the project activities.

Mutual acceptance of the electricity project and its related process in the microgrid communities arose from a participatory stakeholder approach in the implementation of microgrid projects in the communities. This was seen to include the identification and involvement of the relevant stakeholders in the microgrid project, and the readjustment of certain unfavourable implementation terms and conditions to suit their existing implementation context through activities such as lobbying, advocating, and reviewing operation guidelines by the various microgrid stakeholders. Access to electricity connection on a favourable loan instalment basis, restructuring operational regulations to reduce the cost of electricity connection and unit cost, and repealing government policies that sought to drastically cut the cost of microgrid electricity at the expense of microgrid operators' operations are some examples of realignment and mutual acceptance as conveyed via the field interviews. This led to the identification and negotiation of influences that could be detrimental to project implementation; this is what one would call the »balancing of power« as reflected in the Tanzanian microgrid community case below (see Fig. 52).

Balancing Power Framework—Tanzanian Microgrid Community Case

Management structure: Private operator managed
Location: 05 sub-villages
Stakeholders: Operator, local leaders, electricity committee members, Rural Electrification Agency, community users, Energy line ministries
Challenges:

- Limited electricity coverage
- High electricity connection cost
- High per-unit cost
- Exploding meters

Initiation stage: Following extensive and consultative baseline assessments, the microgrid developer/operator held discussions with local community leaders. One village comprised of five sub-villages agreed to receive the solar energy project.

Implementation: Suitable land was located; the school provided an access road, and the operator constructed a playground for the school in exchange. This was followed by an onboarding of local technicians, leaders, and electricity committees. Free electricity connection was provided to the school and the hospital. Mobilization and sensitization of all five sub-villages were carried out by the operator, electricity committee, and village leaders. The operator did not have enough finances and equipment to supply all five sub-villages. The local leaders and the electricity committee successfully lobbied the rural electricity regulatory authority and the Ministry of Energy to subsidize the electricity operator for connection of all five sub-villages.

Monitoring: Connection levels to the electricity project remained low. Consultation meetings with village leaders and electricity users revealed that the connection fee was quite high for the users to pay in one go. Also, the monthly utility bills seemed high for the community that was used to buying daily fuel sources. At one point, continuously exploding meters discouraged users from connecting to the electricity project.

Operation: Development and roll out of favourable instalment pay connection fee terms. Application of daily fee electricity consumption rates that were cheaper than locally used fuels. Productive use of energy (PUE) promotion, periodic multi-stakeholder (operator, community leaders, electricity committee members, community members) engagements for problem identification, resolution, and transparent operations.

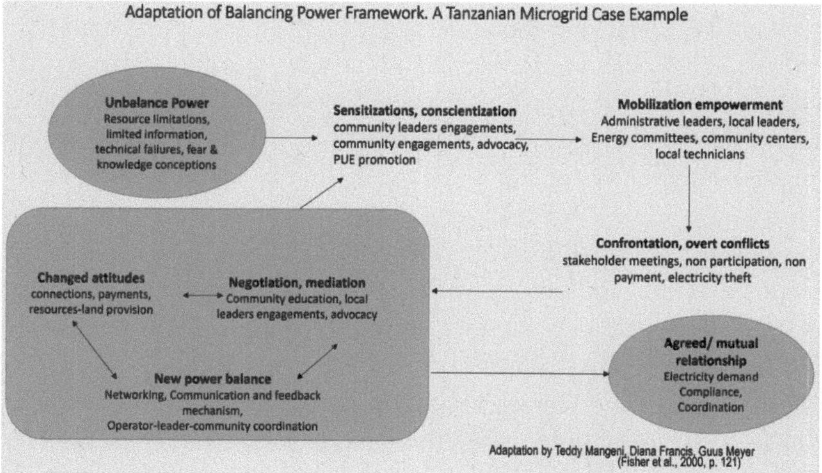

Fig. 52: Adaptation by Mangeni 2024 using field data. Diana Francis, Guus Meyer (Fisher et al., 2000, p. 121).

4.3.5.5 Conclusion

Conflict analysis is a practical process of examining and understanding the reality of the conflict from a variety of perspectives. This understanding then forms the basis on which strategies can be developed and actions planned (Fisher et al., 2000, p. 17). The research conducted in the East African microgrid communities emphasizes the criticality of the fact that beneficiary communities are custodians of their structures, beliefs, and ways. Stakeholder education in energy transitions is not one-sided—the beneficiary community learning from the microgrid developers and their partners opens to the microgrid developers, operators, regulatory authorities, and all outsiders a field to learn, an insight into the community context, and creates a fertile ground to cultivate coordination for sustainable project implementation (Mangeni, 2024, p. 173). If community members cannot accept socio-technological change, they risk losing their willingness to be approached by formal learning. When conflicts occur, they often indicate good reasons for non-acceptance of the microgrid. Instructions on how to use and maintain a microgrid do not seem to address the questions and educational needs of the people, so room is needed, first, to find out about the questions and then find the media to work on answers (Asiimwe et al., 2024, p. 253).

The acceptance and adoption of solar energy projects in the microgrid communities were continually linked to community development. The respondents stated that the electricity projects and their related processes enabled the expansion of existing businesses and initiation of business ventures, longer study hours into the night for school children, easier access to properly stored medicine and medical services, improved and more vibrant social lifestyles, safer electrified night-time communities, enhanced child protection through engaging children at home on television instead of them roaming the streets, longer working hours, and the upscaling of villages to town centre administrative level, among other reported benefits. The field data presents the microgrid community realities and emphasizes that the ability of the electricity projects to contribute to this continuous community development hinges on a meaningful participatory project implementation process.

4.3.6 References

Acquah-Swanzy, H., & Hehenkamp, B. (forthcoming). Nudging and technical interventions in a CPR game with uncertain capacities: A microgrid-framed experimental study.

Africa's Minigrid DevelopersAssociation – AMDA . (2022). *Benchmarking Africa's Minigrid Report.* AMDA. Retrieved from https://africamda.org/wp-content/uploads/2025/01/

Arnstein, S. R. (2019). A ladder of citizen participation. *Journal of the American Planning Association, 85*(1), 24–34. https://doi.org/10.1080/01944363.2018.1559388

Asiimwe, H., Bode, H., Bogere, P., Freitag, C., & Mangeni, T. (2024). Which media for whom? The implementation of microgrids as a trigger of transformational adult learning opportunities in formal, informal and situational settings in times of change. In E. Andrzejewska, E. Matthes, S. Schütze, & J. Van Wiele (Eds.), *Bildungsmedien für Erwachsene / Educational Media for Adults* (pp. 245–255). Verlag Julius Klinkhardt. https://doi.org/10.35468/6126

Asiimwe, H., Freitag, C., & Mangeni, T. (2025). Education as negotiation? Lessons to be learned from the social side(s) of sustainable renewable energy development in East Africa. In: J. Costa et al. (Ed.), *Bildung für nachhaltige Entwicklung: Kontroversen und Debatten.* Verlag Barbara Budrich. https://doi.org/10.3224/84743144

Babayomi, O., & Okharedia, T. (2019). Challenges to sub-Saharan Africa's renewable microgrid expansion—a CETEP solution model. *2019 IEEE PES/IAS PowerAfrica,* 617–621. https://doi.org/10.1109/PowerAfrica.2019.8928865

Batel, S., & Devine-Wright, P. (2015). Towards a better understanding of people's responses to renewable energy technologies: insights from social representations theory. *Public Understanding of Science, 24*(3), 311–325. https://doi.org/10.1177/0963662513514165

Budescu, D. V., Rapoport, A., & Suleiman, R. (1995). Common pool resource dilemmas under uncertainty: qualitative tests of equilibrium solutions. *Games and Economic Behavior, 10*(1), 171–201.

Duran, A. S., & Sahinyazan, F. G. (2021). An analysis of renewable mini-grid projects for rural electrification. *Socio-Economic Planning Sciences, 77*(August 2020), 100999. https://doi.org/10.1016/j.seps.2020.100999

Fisher, S., Ludin, J., Williams, S., Abdi, I., Smith, R., & Williams, S. (2000). *Working with Conflict: Skills and Strategies for Action.* Zed Books.

Fisher, S., Matovic, V., Ludin, J., Ibrahim, D. A., Walker, B. A., Smith, R., Mathews, D., Williams, S., & Williams, S. (2000). Working with Conflict 2: Skills and Strategies for Action.

Galtung, J. (1996). *Peace by Peaceful Means: Peace and Conflict, Development and Civilization.* SAGE Publications Ltd. https://doi.org/10.4135/9781446221631

Gehrig, S., Schlüter, A., & Hammerstein, P. (2019). Sociocultural heterogeneity in a common pool resource dilemma. *PLoS One, 14*(1): e0210561.

Ikejemba, E. C. X., Mpuan, P. B., Schuur, P. C., & Hillegersberg, J. van. (2017). The empirical reality and sustainable management failures of renewable energy projects in sub-Saharan Africa (part 1 of 2). *Renewable Energy, 102*, 234–240. https://doi.org/10.1016/j.renene.2016.10.037

Kakande, J. N., Philipo, G. H., & Krauter, S. (2024) Optimized e-mobility and portable storage integration in an isolated rural solar microgrid in Uganda. *Solar 2024, 4*, 694–727. https://doi.org/10.3390/solar4040033

Kidwai, A. H., & de Oliveira, A. C. (2020). Threshold and group size uncertainty in common-pool resources: an experimental study. *Public Finance Review, 48*(6), 751–777.

Lyamouri-Bajja, N., Genneby, N., Markosyan, R., & Ohana, Y. (2011). *Transforming Conflict*, Vol. 12. Council of Europe.

Maas, A., Goemans, C., Manning, D., Kroll, S., & Brown, T. (2017). Dilemmas, coordination and defection: how uncertain tipping points induce common pool resource destruction. *Games and Economic Behavior, 104*, 760–774.

Maier, C. (2007). Decentralised rural electrification by means of collective action: the sustainability of community-managed micro hydels in Chitral, Pakistan. *Occasional Working Papers Geographie, 33*. https://refubium.fu-berlin.de/bitstream/fub188/18198/1/OccPapers33_Maier.pdf

Mangeni, T. N. (2024). *Negotiating Power Acceptance for Sustainable Minigrid Project Implementation*. Universität Bibliothek Paderborn. https://doi.org/10.17619/UNIPB/1-2144

Mwammenywa, I., Kagarura, G. M., Petrov, D., Holle, P., & Hilleringmann, U. (2021). LoRa-based demand-side load monitoring and management system for microgrids in Africa. *International Conference on Electrical, Computer, and Energy Technologies, ICECET 2021, December*, 1–5. https://doi.org/10.1109/ICECET52533.2021.9698506

Mwammenywa, I., Petrov, D., Holle, P., & Hilleringmann, U. (2022). LoRa transceiver for load monitoring and control system in microgrids. *8th International Conference on Engineering and Emerging Technologies, ICEET 2022, October*, 1–5. https://doi.org/10.1109/ICEET56468.2022.10007274

Namujju, L. D., Mwammenywa, I., Kagarura, G. M., Hilleringmann, U., & Hehenkamp, B. (2024). Smart metering and choice architecture in demand-side management: a power resource-constrained perspective. *2024 IEEE 8th Energy Conference (ENERGYCON)*, Doha, Qatar, 1–6, https://doi:10.1109/ENERGYCON58629.2024.10488738

Ngoti, I. F. (2024). The role of sense of ownership in rural community mini-grid management: qualitative case study from Tanzania. *Energy, Sustainability and Society, 14*(1), 63. https://doi.org/10.1186/s13705-024-00496-7

Odarno, L., Sawe, E., Swai, M., Katyega, M. J. J., & Lee, A. (2017). *Accelerating Mini-Grid Deployment in Sub-Saharan Africa: Lessons from Tanzania*. World Resources Institute. https://files.wri.org/d8/s3fs-public/accelerating-mini-grid-deployment-sub-saharan-africa_1.pdf

Ogeya, M., & Lambe, F. (2025). The political economy of mini-grid electricity development and innovation in Kenya. *Renewable and Sustainable Energy Transition, 6*, 100092. https://doi.org/10.1016/j.rset.2024.100092

Palit, D., & Chaurey, A. (2011). Off-grid rural electrification experiences from South Asia: status and best practices. *Energy for Sustainable Development, 15*(3), 266–276. https://doi.org/10.1016/j.esd.2011.07.004

Mini-Grids Partnership. (2020). *State of the Global Mini-grids Market Report 2020: Trends of Renewable Energy Hybrid Mini-grids in Sub-Saharan Africa, Asia and Island Nations.* Mini-Grids Partnership (MGP), BloombergNEF, Sustainable Energy for All. https://www.seforall.org/system/files/2020-06/MGP-2020-SEforALL.pdf

Peters, J., Sievert, M., & Toman, M. A. (2019). Rural electrification through mini-grids: challenges ahead. *Energy Policy, 132*(April), 27–31. https://doi.org/10.1016/j.enpol.2019.05.016

Polatidis, H., & Haralambopoulos, D. A. (2004). Local renewable energy planning: a participatory multi-criteria approach. *Energy Sources, 26*(13), 1253–1264. https://doi.org/10.1080/00908310490441584

Rapoport, A., & Au, W. T. (2001). Bonus and penalty in common pool resource dilemmas under uncertainty. *Organizational Behavior and Human Decision Processes, 85*(1), 135–165.

Travers, H., Clements, T., Keane, A., & Milner-Gulland, E. (2011). Incentives for cooperation: the effects of institutional controls on common pool resource extraction in Cambodia. *Ecological Economics, 71*, 151–161.

Tenenbaum, B., Greacen, C., Siyambalapitiya, T., & Knuckles, J. (2014). From the bottom up: how small power producers and mini-grids can deliver electrification and renewable energy in Africa. *Landscape Architecture, 93*(9). https://doi.org/10.7146/chku.v7i1.138103

Uvin, P. (1996). Development, aid and conflict reflections from the case of Rwanda. *WIDER Research for Action*, Issue 24. UNU World Institute for Development Economics Research.

Ware, A., & Laoutides, C. (2021). Whose analysis? Trial of a new participatory conflict analysis for do no harm/conflict-sensitive development planning. *Conflict, Security and Development, 21*(5), 673–696. https://doi.org/10.1080/14678802.2021.1984682

Zigah, E., Barry, M., & Creti, A. (2023). Are mini-grid projects in Tanzania financially sustainable? In S. Groh, L. Barner, G. Heinemann, & C. Von Hirschhausen (Eds.), *Electricity Access, Decarbonization, and Integration of Renewables* (pp. 233–261). Springer Fachmedien Wiesbaden. https://doi.org/10.1007/978-3-658-38215-5_10

4.4 Gender Discourses in Energy Development

Henry Asiimwe

4.4.1 Introduction

Energy development is a cornerstone of global progress, shaping economies, societies, and individual lives. From enabling industrial growth to providing essential services like lighting, heating, and communication, energy is a driver of economic development and human well-being. However, its benefits and burdens are unevenly distributed across populations, often reinforcing pre-existing inequalities. One of the critical dimensions of this disparity is gender—a factor that profoundly influences access, participation, and decision-making in energy development processes (Clancy et al., 2003, p. 356; Köhlin et al., 2011, p. 1260). Despite growing recognition of gender's role in sustainable development, the interplay between energy systems and gender remains underexplored and poorly integrated into policy and practice (Rao & Reddy, 2007, pp. 143–153; Pachauri & Rao, 2013, p. 16).

Energy access is a key component of achieving equity and social justice, yet millions of women worldwide remain marginalized in energy systems. Women, particularly in low- and middle-income countries, are disproportionately affected by energy poverty, relying heavily on traditional biomass for cooking and heating (Batliwala & Reddy, 2003, p. 375; IEA, 2017, p. 35). This reliance has cascading effects on their daily lives, from spending hours collecting fuelwood to facing increased risks of respiratory diseases due to indoor air pollution (Parikh, 2011, p. 248; WHO, 2018, p. 41). These challenges are further exacerbated by the gendered division of labour, where women often bear the primary responsibility for household energy needs. Consequently, energy poverty not only limits women's economic and educational opportunities but also perpetuates intergenerational cycles of poverty.

In contrast, the energy sector remains largely male-dominated, both in technical fields and in decision-making positions (Cecelski, 2000; Clancy et al., 2012, p. 27). This exclusion of women from the energy development process has profound implications for the effectiveness and inclusivity of energy policies and programmes. For example, women's unique knowledge and perspectives, particularly regarding household energy usage, are often overlooked in the design of energy solutions. Moreover, the exclusion reinforces systemic barriers that prevent women from participating as entrepreneurs, innovators, and leaders in the energy sector (Oparaocha & Dutta, 2011, p. 45). Addressing this imbalance is not merely a question of representation; it is a critical factor for ensuring equitable and sustainable energy systems.

This article critically examines the dominant gender discourses in energy development, exploring how these narratives shape policies, practices, and outcomes. It highlights the evolution of gendered perspectives in energy, from their marginal inclusion in early energy access discussions to their growing prominence in global agendas, such as the Sustainable Development Goals (SDGs) (United Nations, 2015, p. 14; Byrne et al., 2014, p. 14). These global commitments emphasize the importance of gender equality as a cross-cutting issue, recognizing that energy systems cannot be sustainable without addressing the gendered disparities within them.

However, despite these advancements, challenges remain. Gender discourses in energy often risk being tokenistic, focusing on inclusion as a checkbox exercise rather than addressing underlying structural inequalities (Clancy et al., 2012, p. 27; Winther et al., 2017, p. 389). For example, many energy projects continue to prioritize technical and economic efficiency over social inclusivity, side-lining the voices of women and marginalized groups. This approach not only undermines the transformative potential of energy interventions but also risks perpetuating the very inequalities these projects seek to address.

The article argues that understanding gender in energy development requires moving beyond simplistic notions of women as passive beneficiaries. Women are not merely recipients of energy interventions; they are active agents of change who play diverse roles as energy users, entrepreneurs, and leaders. For instance, in many rural communities, women-led initiatives have demonstrated the potential of renewable energy technologies to empower communities and improve livelihoods (Skutsch, 2005, p. 336).

Recognizing these contributions is essential for designing energy policies and programmes that are both inclusive and effective.

Moreover, addressing gender disparities in energy development necessitates a holistic approach that considers the intersectionality of gender with other factors such as class, ethnicity, and geographic location (Crenshaw, 1991, p. 1245). Women in urban areas, for example, face different energy challenges than those in rural settings, just as women from marginalized ethnic groups may experience unique forms of exclusion. Policymakers must take these nuances into account to ensure that energy interventions are responsive to the diverse needs and realities of women across different contexts.

Finally, the article underscores the importance of fostering an enabling environment for gender equality in the energy sector. This includes investing in education and capacity-building programmes to empower women as energy professionals, promoting gender-responsive financing mechanisms, and creating platforms for women's voices in decision-making processes (Winther et al., 2017, p. 389). By doing so, the energy sector can not only achieve greater equity but also unlock innovative solutions to some of its most pressing challenges.

Ultimately, the article positions gender as a critical lens through which energy development must be understood. It invites policymakers, practitioners, and researchers to rethink traditional approaches, advocating for a shift from technocratic solutions to those that are socially inclusive and culturally responsive. Energy systems must evolve to be not only efficient and resilient but also equitable and empowering, contributing to a future where all individuals have the opportunity to thrive. By embracing gender as a central consideration, the energy sector can become a catalyst for transformative social and economic change.

4.4.2 Literature on Gender Discourses and Barriers to Equitable Participation in Renewable Energy Processes

Achieving gender equity in the renewable energy sector is crucial for fostering sustainable energy systems and promoting socio-economic development. Although awareness of this issue has grown, pervasive gendered narratives continue to obstruct women's participation in transformational enterprises, especially in renewable energy initiatives. This review explores how such gendered discourses influence women's involvement in renewable energy projects, with a specific focus on microgrids in East Africa. These narratives not only curtail women's access to opportunities in the sector but also uphold systemic power structures that limit their active engagement. By analysing existing literature, this review highlights the challenges presented by these discourses and explores the transformative potential of inclusive energy policies and practices.

The importance of gender equity in renewable energy has been extensively documented. Women's involvement in managing energy resources and contributing to decision-making processes significantly improves the sustainability and adaptability of energy systems. Research by Baruah (2017, p. 72) highlights that including women enhances resource efficiency and energy conservation. Supporting this, Clancy et al. (2019, p. 33) argue that women's perspectives are crucial in understanding community energy needs and designing responsive projects. Global organizations like the International Energy Agency (IEA, 2022, p. 15) underscore that a gender-diverse energy workforce fosters innovation and produces better outcomes, making it essential for meeting international sustainability objectives. Nonetheless, women remain underrepresented in the renewable energy sector, particularly in East Africa, where cultural norms and gendered practices systematically marginalize them (Graber & Moksnes, 2020, p. 89).

Gendered narratives that restrict women's participation in the energy sector are deeply rooted in societal norms and traditions. In East African contexts, these narratives often portray energy-related tasks, such as installing or maintaining microgrids, as unsuitable for women (IRENA, 2019, p. 45). Activities like climbing poles, operating equipment, and handling machinery are typically framed as male domains, reinforcing stereotypes about women's technical ineptitude. According to Michael and Ahlborg (2024, p. 951), these stereotypes perpetuate exclusion and discourage women from pursuing careers in technical fields. Consequently, women are frequently excluded from both technical and leadership roles in energy projects, especially those related to microgrid development. These patterns of exclusion reflect broader systemic inequities. According to IRENA (2019, p. 48), the absence of gender-sensitive energy policies perpetuates the underrepresentation of women in decision-making and technical positions. Gender-blind energy policies worsen this disparity by failing to address the unique obstacles women face in accessing education, training, and employment opportunities within the energy sector.

Despite these barriers, the transformative potential of renewable energy projects for women is increasingly being recognized. Electrification through microgrids, in particular, can alleviate the »triple burden« many women face, managing reproductive, productive, and community roles, thereby enabling them to pursue income-generating activities and leadership opportunities. Studies by Allen et al. (2020, p. 102) reveal that access to energy empowers women by enhancing their negotiation abilities at household and community levels, allowing them to play more significant roles in decision-making. Furthermore, projects that actively involve women often yield broader benefits for entire communities, including better health, education, and economic outcomes (Clancy et al., 2019, p. 36).

Addressing these barriers requires targeted policies and interventions. Scholars like Baruah (2017, p. 76) advocate for gender-sensitive energy policies designed to dismantle cultural and societal barriers that restrict women's access to opportunities in the renewable energy sector. Michael and Ahlborg (2024, p. 948) emphasize that such policies must also include targeted training initiatives to build women's technical expertise and leadership capacity. These interventions empower women to contribute meaningfully to energy projects, ultimately improving the efficiency and sustainability of these efforts.

In addition to policy changes, community-driven initiatives are essential for reshaping socio-cultural norms and challenging the gendered narratives that exclude women from the energy sector. Collaborating with local stakeholders and community leaders to raise awareness about gender equity can help dismantle stereotypes and foster greater acceptance of women in technical and leadership roles. By emphasizing the community-wide benefits of women's participation in renewable energy projects, such initiatives can create more inclusive and equitable energy systems (Graber & Moksnes, 2020, p. 94).

4.4.3 Local Discourses on Women's Participation: Insights from East Africa

Bishoge et al. assert that like any initiative aimed at empowering communities, inclusive participation of all groups is essential for sustainability (2020, p. 45). Following this line of thought, for projects like renewable energy microgrids to succeed, women's involvement and considerations of gender equity are critical (Fraune, 2015, p. 55). Hence, this study places particular emphasis on women because, as scholars like Pedersen highlight, the lack of electricity and other energy resources in rural East Africa disproportionately impacts them compared to other community members (Pedersen, 2016, p. 73). Despite their active involvement, women's participation reflects a duality of supportive discourses and discursive barriers, as explored in this section. The discussions in the proceeding section are based on empirical research conducted in East Africa with primary data collected from Uganda and Tanzania from 2021 to 2024. The study relied on discourse analysis as a methodological approach and collected data from 16 microgrid energy communities targeting energy users, local leaders, investors in the microgrids energy sector, and government leaders working in departments related to and dealing with energy and gender sectors. The results revealed a host of local discourses that perpetuate gendered participation in the energy sector. A few of them are highlighted in the subsections that follow.

4.4.3.1 Cooperative Discourses for Women's Participation in Energy Processes

Although much literature highlights the challenges hindering women's involvement in renewable energy and technology, this study uncovers evidence of women's significant contributions to microgrid development. Nonetheless, structural obstacles persist. For instance, microgrid projects often start with feasibility studies and community consultations. During these engagements, women formed a considerable part of those whose opinions were solicited. While their input on electricity needs and productive uses was valued, gender biases surfaced during household surveys. Data collectors typically consulted men when both spouses were present, and even in women's absence, men's names were often recorded. This reflects entrenched notions of men as household heads, a belief deeply ingrained in many African and Asian cultures (Bishoge et al., 2020, p. 47). Despite these biases, some women hold administrative positions in microgrid projects, managing sales, marketing, communication, and customer care. Out of 11 microgrids visited, 8 had women administrators acting as key community representatives. However, technical roles remained largely male-dominated, reflecting a gendered labour division consistent with broader studies of gender dynamics in the energy sector (Fraune, 2015, p. 56). One administrator remarked that women excel in customer care due to their patience and ability to manage dissatisfied customers, particularly in disorganized rural settings with high tension. While this belief is partly rooted in stereotypes, it has helped women secure positions in this traditionally male-dominated sector (Clancy et al., 2012, p. 16).

In addition to administrative duties, women serve on local energy management committees, undertaking responsibilities such as customer mobilization, solar panel cleaning, and enforcing tariff payments. In some sites, like KS in central Uganda, women comprised the entire mobilization committee, significantly increasing customer connections. In Tanzania's SL microgrid, a female chairperson successfully led the energy committee, a role historically reserved for men (Pedersen, 2016, p. 77).

Women's attendance at community meetings and energy-related training sessions was also noteworthy, often exceeding that of men due to their strong interest in community development. A local leader observed that women's productive use enterprises expanded more rapidly than those run by men because women were more attentive and proactive in seeking knowledge. Contrary to assumptions that African women remain silent in public gatherings due to patriarchal cultural norms, women in several communities voiced their opinions actively, even in mixed-gender settings. However, cultural norms still restrict full participation. Women are expected to return home before dark, limiting their involvement in evening activities. This double standard reflects the power dynamics that grant men greater freedoms while constraining women's movements (Clancy et al., 2012, p. 17). These norms are so ingrained that even when women's participation would benefit the community, such as in energy security committees, men resist their involvement for fear of social judgement. One focus group respondent remarked that walking with a woman at night could raise suspicions of an affair, discouraging men from including women in night-time security tasks.

Despite these challenges, women have made notable progress in technical roles. During a microgrid construction in Uganda, female electrical engineers trained at universities worked alongside men to install the grid, a rare occurrence in a sector dominated by male technicians. Outside microgrids, more women are becoming electricians, gradually dismantling gender stereotypes in technology and energy fields (Clancy et al., 2012, p. 19). Nevertheless, stereotypes persist, creating obstacles for women in technical roles. A male manager at a Ugandan microgrid company admitted that they rarely hire women for technical tasks, perceiving the roles as being too physically demanding. Even when women participated in solar plant setups, their roles were often confined to cleaning rather than machinery operation. This reflects a widespread perception that women are unsuited for engineering and technical fields despite evidence to the contrary (Bishoge et al., 2020, p. 51).

It was also argued by participants that when women are given opportunities, they frequently outperform men in terms of dedication, honesty, and diligence. Several respondents noted that women approach their work with greater enthusiasm, driven by the need to support their families, especially in cases where men neglect these responsibilities. In some communities, local leaders observed that women's businesses thrive because they invest more time and effort into learning and applying new knowledge, unlike men who often resist change or view meetings as unproductive, as claimed by one local village leader in Central Uganda.

A female customer care worker at a microgrid site in central Uganda commented that women's ability to multitask and manage household needs makes them particularly adept at roles in the energy sector. She noted that women's empathy towards customers facing power outages leads to better service and satisfaction. Conversely, male technicians, under pressure from polygamous family dynamics and social expectations, were less concerned with promptly resolving customer issues. Such observations illustrate the deeply rooted discourses supporting and opposing women's participation.

4.4.3.2 Discourses Reinforcing Gender Stereotypes and Gendered Participation in Energy

Cultural norms and gender stereotypes strongly shape societal expectations about the roles men and women occupy, particularly regarding work (Clancy et al., 2012, p. 23). In many Ugandan communities, some traditions establish rigid divisions between »men's work« and »women's work«. Women are often confined to roles perceived as less physically or intellectually demanding, limiting their access to technical professions like engineering or energy-related fields. Findings show that Uganda's energy sector is predominantly seen as a male domain due to entrenched cultural beliefs that technical roles require strength and intellectual capacity—qualities that women, according to local discourses, are presumed to lack. These perceptions deter women from pursuing careers in these fields, regardless of their abilities or qualifications. In rural areas, strict gender boundaries further constrain women. Instances were noted where women entering male-dominated roles, such as solar panel installation, faced ostracism and derogatory labels. Cultural taboos portray such women as transgressing societal norms, threatening the social order. In extreme cases, some communities viewed women's involvement in male roles as bringing bad luck or being a »bad omen«, reinforcing their exclusion. Additionally, cultural norms emphasizing physical strength as a masculine trait further entrench the belief that women cannot handle physically demanding tasks (Bishoge et al., 2020, p. 55). These stereotypes, perpetuated through daily interactions, bolster the status quo by associating physical capability with technical competence.

Religion also plays a role in reinforcing stereotypes. Dominant Christian teachings in Uganda often reference biblical texts portraying women as subordinate to men. For example, biblical accounts of Eve's creation from Adam's rib are frequently cited to justify women's perceived inferiority. In Uganda's deeply religious society, such teachings significantly influence gender roles and expectations. During focus group discussions, such arguments prominently cropped up as reasons why women must remain subordinate to men, and such being subordinate extended to refraining from challenging decisions from male partners regarding energy investments.

Socialization from childhood perpetuates these stereotypes. Boys and girls are raised with distinct societal roles, instilling ideas of gender inequality (Fraune, 2015, p. 63). From a young age, boys are groomed for leadership and technical work, while girls are encouraged to embrace nurturing and domestic tasks. This early socialization limits women's career aspirations and reinforces the notion that technical fields are inappropriate for them. For instance, in central Uganda, young girls undergo formal training to learn domestic roles, while boys are taught skills for technical and leadership positions (Witt, 1997, p. 253). Even educated women may steer clear of technical professions to avoid societal backlash. Ridicule and exclusion serve as potent tools for enforcing conformity, discouraging women from challenging traditional norms (Clancy et al., 2012, p. 27).

4.4.3.3 Household Power Negotiations and Decision-Making

Household power relations mirror societal patriarchy, where men predominantly hold decision-making authority (Pedersen, 2016, p. 87). In East African societies, these dynamics permeate economic, career, and energy-related decisions. Men often control household decisions on electricity use, appliance purchases, and their wives' and daughters' career choices. This study highlights examples where women's participation in energy committees or employment in technical fields depended on their husbands' approval. Such restrictions limit women's professional opportunities, reinforcing their subordinate household roles (Witt, 1997). Cultural and religious beliefs support these power imbalances. In many Ugandan households, male dominance is viewed as natural and divinely ordained. Focus group discussions revealed that older women endorsed men's veto power over household decisions, citing religious teachings. Biblical passages describing men as household heads and women as submissive reinforce these dynamics, making it difficult for women to challenge male authority (Ellemers, 2018, p. 275).

Exclusion from household energy decisions reflects broader societal beliefs about women's perceived inability to handle technical matters. However, the research also found that when women earn independent incomes, household power dynamics begin to shift. Women's financial contributions often lead to greater bargaining power, though this shift can provoke resistance from men attempting to reassert control by restricting access to income or influencing its use. These tensions illustrate the fragile balance of traditional authority amid women's growing economic empowerment (Fraune, 2015, p. 67).

4.4.3.4 Discourses Commodifying the Female Body

Societal discourses surrounding women's bodies impose significant restrictions on their participation in technical fields. In many African communities, women's bodies are objectified and regulated by strict social norms (Clancy et al., 2012, p. 31). A prominent belief was revealed by a number of respondents that certain physical activities, such as climbing poles, compromise a woman's virginity, which is highly valued for her social and marital prospects. Such beliefs deter young women from pursuing careers requiring these tasks, fearing familial shame and diminished marriage opportunities. Clothing norms were also found to further reinforce gender barriers. Women are expected to dress modestly, often in long skirts or dresses, which are impractical for technical work like electrical engineering. It was revealed that wearing safety gear, such as trousers, is discouraged as »unfeminine«. These norms prevent women from entering professions requiring practical attire, reinforcing their exclusion from technical fields (Fraune, 2015, p. 69). Moreover, such societal objectification reduces women's value to their appearance. For instance, businesses in urban areas were found to often prefer hiring women to attract customers based on looks rather than skill. This commodification marginalizes women from technical professions demanding expertise and physical labour. Such barriers further perpetuate broader gender inequalities, limiting women's career opportunities and reinforcing societal biases (Bishoge et al., 2020, p. 65).

4.4.4 Conclusion

This section has explored the intricate and often contradictory gender discourses that shape women's participation in energy development, with a particular focus on renewable energy initiatives in East Africa. Through an intersectional lens and grounded in empirical data, it has shown how dominant narratives continue to marginalize women by reinforcing gendered stereotypes, cultural norms, and institutional barriers. These discourses frequently position technical and leadership roles in the energy sector as male domains, while assigning women supportive or administrative roles. In many instances, societal beliefs, religious teachings, and household power dynamics further entrench women's exclusion from energy-related decision-making and technical employment.

Yet, the analysis also reveals a more hopeful dimension. Despite these constraints, women in several microgrid projects have actively participated in customer care, mobilization, and even technical installations. They have demonstrated innovation, diligence, and leadership—traits that are touted as crucial for the success of community energy projects. These positive examples underscore the transformative potential of inclusive energy systems when deliberate efforts are made to meaningfully include all categories of people.

However, realizing this potential requires a fundamental shift in how gender is understood and addressed in energy development. Tokenistic approaches must give way to strategies that confront structural inequalities and empower women as co-creators of energy solutions. Gender-sensitive policies could incorporate training, mentorship, and financing mechanisms that address specific barriers faced by women. At the same time, community engagement should aim to shift harmful gender norms by working with both men and women to redefine roles and expectations.

From the foregoing, sustainable energy transitions cannot be achieved without addressing the gendered dimensions of access, participation, and power. Women are not merely passive beneficiaries but key stakeholders whose contributions can drive equitable and resilient energy futures. Embracing inclusive gender discourses is therefore not just optional but is essential for social justice and sustainable energy development.

4.4.5 References

Allen, C., Smith, J., & Njeri, P. (2020). Energy Access and Gender Empowerment: Pathways for Change in Sub-Saharan Africa. Oxford University Press.

Baruah, B. (2017). Renewable Inequities: Women's Role in the Transition to Renewable Energy. Routledge.

Batliwala, S., & Reddy, A. K. N. (2003). Energy for women and women for energy: empowering women through energy entrepreneurship. Energy for Sustainable Development, 7(3), 367–382.

Bishoge, O., Kombe, G., & Mvile, B. (2020). Community participation in the renewable energy sector in Tanzania: community participation in the renewable energy sector. International Journal of Sustainable Energy Planning and Management, 28, 121–134.

Byrne, R., Smith, A., Watson, J., & Ockwell, D. (2014). Energy pathways in low-carbon development: from technology transfer to socio-technical transformation. Energy Policy, 48, 12–20.

Cecelski, E. (2000). The Role of Women in Sustainable Energy Development. National Renewable Energy Laboratory. https://docs.nrel.gov/docs/fy00osti/26889.pdf.

Clancy, J. S., Oparaocha, S., & Roehr, U. (2012). Gender equity and renewable energies. Thematic Background Paper, International Conference on Renewable Energies, 21–33.

Clancy, J. S., Skutsch, M., & Batchelor, S. (2003). The gender-energy-poverty nexus: can we find the energy to address gender concerns in development? Energy Policy, 31(11), 355–367.

Clancy, J. S., Skutsch, M., & Batchelor, S. (2019). Energy, Gender, and Development: What is the Link? Earthscan Publications.

Crenshaw, K. W. (1991). Mapping the margins: intersectionality, identity politics, and violence against women of color. Stanford Law Review, 43(6), 1241–1299.

Ellemers, N. (2018). Gender stereotypes. Annual Review of Psychology, 69, 275–276.

Fraune, C. (2015). Gender matters: women, renewable energy and citizen participation in Germany. Energy Research and Social Science, 7, 55–65.

Graber, T., & Moksnes, H. (2020). Gender Equality in Energy Access: Lessons from East Africa. Palgrave Macmillan.

International Energy Agency (IEA). (2017). Energy Access Outlook 2017: From Poverty to Prosperity. IEA.

International Energy Agency (IEA). (2022). Energy Policies for Gender Equality: A Global Perspective. IEA.

International Renewable Energy Agency (IRENA). (2019). Gender and Renewable Energy: Empowering Women through Access to Energy. IRENA.

Köhlin, G., Sills, E. O., Pattanayak, S. K., & Wilfong, C. (2011). Energy, gender and development: what are the linkages? where is the evidence? World Bank Policy Research Working Paper, 5800, 1240–1275.

Michael, K., & Ahlborg, H. (2024). A conceptual analysis of gendered energy care work and epistemic injustice through a case study of Zanzibar's Solar Mamas. Nature Energy, 9, 947–954. https://doi.org/10.1038/s41560-024-01539-1.

Oparaocha, S., & Dutta, S. (2011). Gender and energy for sustainable development. Current Opinion in Environmental Sustainability, 3(4), 265–271.

Pachauri, S., & Rao, N. D. (2013). Gender impacts and determinants of energy poverty. Energy Policy, 56, 12–23.

Parikh, J. (2011). Gender and sustainable energy: issues and opportunities. Energy Policy, 39(5), 245–255.

Pedersen, M. B. (2016). Deconstructing the concept of renewable energy-based minigrids for rural electrification in East Africa. Wiley Interdisciplinary Reviews: Energy and Environment, 5(5), 570–587. https://doi.org/10.1002/wene.205

Rao, N., & Reddy, B. S. (2007). Variations in energy use by Indian households: an analysis of micro level data. Energy, 32, 143–153. https://doi.org/10.1016/j.energy.2006.03.012

Skutsch, M. M. (2005). Gender analysis for energy projects and programs. Energy for Sustainable Development, 9(1), 333–343.

United Nations. (2015). Transforming Our world: The 2030 Agenda for Sustainable Development. United Nations.

Winther, T., Matinga, M. N., Ulsrud, K., & Standal, K. (2017). Women's empowerment through electricity access: scoping study and implications for research. Journal of Development Effectiveness, 9(3), 389–417.

Witt, S. (1997). Parental influence on children's gender roles. Adolescence, 32(126), 253–259.

World Health Organization (WHO). (2018). Household Air Pollution and Health. WHO.

5. Educational Perspectives

5.1 Education for Sustainable Development: East African Perspectives

James Musana

5.1.1 Introduction

Eastern Africa, also referred to as East Africa, is a diverse and dynamic region located at the eastern edge of the African continent. It encompasses a variety of countries, cultures, and geographical features, making it a significant area both historically and in contemporary contexts. The countries located in East Africa face similar economic, social, and environmental challenges, such as inequality, high poverty rates, unemployment, unsustainable energy production, and resource degradation (UNECA, 2021).

East African countries have endeavoured to achieve a common subregional integration agenda through the establishment of regional economic communities (RECs) andintergovernmental organizations (IGOs) to foster trade and development (United Nations, 2024). There is cooperation among East African member states towards the development and implementation of some collective programmes which seek to address the challenges and pressures undermining sustainable development (SD) in the subregion (EAC Secretariat, 2022). Some of these programmes—such as the Nile basin initiative programme, conservation and management of natural capital, Water, Sanitation and Hygiene (WASH) and health promotion programme, and climate change adaptation and mitigation programme, among others – are implemented in partnership with development partners. Particularly pertinent to this research, the A:RT-D Grids project has been designed and implemented through cooperation between some universities in East Africa and development partners in Europe to address sustainability and knowledge creation concerns of the renewable energy field in East Africa.

Education for Sustainable Development (ESD) is one of the critical approaches employed in the East African region to address local and global challenges, particularly in relation to the Sustainable Development Goals (SDGs). ESD integrates the inherent values of sustainable development into all aspects and levels of education in order to improve the quality of life of present and future generations. It informs behaviour and lifestyle choices in the three pillars of sustainability by developing knowledge, skills, understanding, values, and actions required to create a sustainable world. ESD incorporates critical SD issues in teaching and learning with local and global perspectives and empowers learners to understand and respond to the changing world. It encourages interdisciplinary and cross-curricular work, strengthens research skills, questioning, risk-taking, critical thinking, and collaborative learning (Annan-diab & Molinari, 2017). ESD also encourages different disciplines to enter into dialogue, make connections, share knowledge, and work together on various issues (The Higher Education Academy Group, 2014). ESD also attempts to address other gaps like gender in education. Integrating gender in ESD is essential for fostering equitable and inclusive societies. By addressing gender disparities, ESD empowers all learners regardless of gender to access quality education and actively participate in sustainable practices. It emphasizes the importance of challenging stereotypes, promoting gender equality, and ensuring that diverse perspectives shape solutions to environmental, social, and economic challenges. This approach not only strengthens individual agency but also enhances collective efforts towards achieving more just and sustainable project outcomes.

It is in this same spirit that the A:RT-D Grids project brought together members of various academic disciplines and continents to examine and implement the project in East African rural microgrid communities. The sustainable energy development project, in the East Africa renewable energy context, employed interdisciplinary approaches to research and project implementation. The project addressed one of the most serious impediments to East African development in a profoundly interdisciplinary and innovative way. The dynamic partnership between the participating academic institutions was enhanced by joint research projects at postgraduate and post-doctoral levels, respectively. Furthermore, ESD puts emphasis on the interconnectedness of the social, economic, cultural, political, and environmental dimensions of SD. This is intended to equip learners with the knowledge and skills necessary for sustainable living. The integration of ESD into educational frameworks in East Africa, therefore, aims at providing learners with the knowledge and skills necessary to promote sustainable practices and contribute positively to their communities. In the East African region, there are various key initiatives, strategies, and approaches which are used to integrate ESD and SD practices. Below are some of those strategies and approaches.

5.1.2 Key Initiatives and Strategies for ESD in East Africa

The East African Community of Practice for Sustainable Development Goal 4 (SDG 4)—to ensure inclusive and equitable quality learning and promote lifelong learning for all—was implemented by UNESCO National Commissions across six East African countries. It focused on collaborative networking to enhance ESD practices. Its aim was to share good practices, harmonize approaches, and develop projects that align with SDG 4, which emphasizes quality education and lifelong learning opportunities for all.

Other strategies have been designed at national level such as the strategy for implementation of ESD in Uganda. The development of ESD strategy involved the participation of a cross-section of various stakeholders in education, economy, environment, and society through meetings, workshops, and seminars (Uganda National Commission for UNESCO, 2010). The strategy gives guidance on how ESD should be implemented, and it considers the national aspirations and goals as stipulated in the Constitution, country's vision 2040, and the National Development Plans. At the end of 2016, a new national ESD policy was enacted, and it provides a framework for reorienting ESD in both formal and informal education (Uppsala University, 2017). The new ESD policy considers not only the national SD aspirations but also regional and international development agendas such as the East African Community Vision 2050, the Africa Union Agenda 2063, and the global agenda 2030 of SDGs (The Republic of Uganda, 2017). In the context of Uganda as stipulated in both the old and the new ESD strategies, the major aim of ESD is to promote values and practical processes which provide knowledge, skills, attitudes, and competences to nurture SD at all levels (Michelsen & Wells, 2017). The Ministry of Education and Sports is the leading ministry in matters concerning ESD and works in partnership with other government regulatory bodies, agencies, education institutions, politicians, civil society, and all other stakeholders in matters of education. Educational institutions are recognized as key players and change agents in instilling and propagating positive attitudes, promoting public awareness and understanding of ESD. On the other hand, Kenya has also developed a national ESD strategy that outlines action-oriented strategies for stakeholders to foster SD. This includes advocacy, capacity-building, and the use of Information and Communication Technologies (ICTs) to enhance educational outcomes. The strategy emphasizes the role of higher education institutions in promoting sustainability through teaching, research, and community engagement. The National Environment Management Authority (NEMA) of Kenya has also established Regional Centres of Expertise (RCEs) that facilitate local ESD initiatives by engaging various stakeholders at different levels. These RCEs promote consultation and participation in planning and implementing sustainability education programmes. This collaborative framework helps to ensure that diverse perspectives are considered in the planning and implementation of sustainability education (Mochizuki & Fadeeva, 2010).

The national frameworks for ESD are essential in promoting SD through education. By aligning educational strategies with global sustainability goals, East African countries can foster a more informed and engaged populace capable of addressing pressing global challenges (The East African Community, 2015). The collaborative efforts among governments, educational institutions, and communities are vital in realizing the potential of ESD as a transformative force for sustainable futures.

Besides the development of strategies at the national level, there are also regional collaborations. Various partnerships have been established to enhance ESD across the region. For instance, the ESD in East Africa project brings together multiple universities working to integrate sustainability into their curricula and community outreach programmes. The involvement of various universities and other stakeholders has helped to record substantial ESD-related practices, which provides an opportunity for expanded ESD mainstreaming in East Africa. The East African Community of Practice for SDG 4 includes stakeholders from Kenya and Uganda working together to share knowledge and best practices related to ESD (Agoi & Nakato, 2012). This initiative fosters collaboration among UNESCO National Commissions and local NGOs focused on environmental education. The community aims to harmonize approaches to ESD across the region. These various initiatives are informed by the various perspectives elaborated below.

5.1.3 Perspectives of Education for Sustainable Development in East Africa

Sustainability issues are complex and context-specific; therefore, various perspectives can be adapted and adopted for the integration of ESD. ESD perspectives in this study are particular ways in which ESD is considered and regarded in East Africa. This can be traced from the historical, political, economic, environmental, social, and cultural background of the people in the region. How one perceives an event can significantly affect one's understanding and response to it. For example, viewing a conflict from multiple perspectives can lead to a more nuanced understanding of the issues involved (Carton & Tewfik, 2016; Omisore & Abiodun, 2014). In East Africa, ESD has been shaped by various perspectives that reflect the region's unique social, economic, and environmental contexts (Tikly, 2019). Some of the key perspectives on ESD, which have been highlighted from recent literature in relation to ESD in East Africa, are presented below.

One of the key perspectives of ESD in East Africa is contextual relevance. Considering the context is crucial for the integration of ESD because it ensures that educational practices and curricula are aligned with the specific social, economic, and environmental conditions of the communities (Hargreaves, 2008). The context of ESD can be perceived through various dimensions, including its integration into curricula, its role in community engagement, and its alignment with global sustainability goals (Nantale et al., 2022; Oe et al., 2022). ESD serves as a catalyst for community learning and engagement. It encourages collaborative networks that enhance environmental awareness and promote behavioural change within communities. For instance, community learning centres have been identified as crucial spaces for fostering discussions around sustainability issues such as poverty, climate change, gender, resource management, and health behaviours (Green & Green, 2016). In this case, community becomes a resource for learning. These centres not only provide resources but also create a supportive environment for individuals to engage in sustainable practices (Araneo, 2024). By linking education with real-world issues, ESD empowers individuals to take informed actions towards sustainability in their communities (Oe et al., 2022). Furthermore, tailoring ESD initiatives to the socio-economic conditions of the region helps to ensure that they are practical and achievable within the local context. For instance, integrating ESD into curricula helps to develop skills that match local economic needs. This alignment not only prepares students for future employment but also contributes to community resilience by equipping them with the necessary tools to address local challenges (Hargreaves, 2008; UNESCO & UNEVOC, 2010).

Another East African perspective on ESD is the consideration of the local cultural heritage. This perspective emphasizes the importance of integrating local knowledge systems into educational frameworks, thereby enhancing relevance and engagement among learners (Zazu, 2016). Engaging local communities in the identification, design, and implementation of educational programmes fosters a sense of ownership and relevance. When students see how their education directly relates to their community's challenges, this enhances their motivation and commitment to sustainable practices (Uganda National Commission for UNESCO, 2010). This approach helps students connect with their heritage while learning about sustainability (Zazu, 2016). The cultural heritage serves as a bridge between past and future, preserving collective memory and providing a foundation for societal development (Tikly, 2019). It is recognized as an integral part of people's cultural identity and plays a significant role in SD by promoting social cohesion. Incorporating local cultural heritage into ESD is essential for fostering sustainable practices and enhancing community identity (Green & Green, 2016). This integration not only enriches educational experiences but also aligns with broader sustainability goals. Incorporating local cultural heritage into ESD enriches, therefore, the learning experience by connecting students with their community's values and traditions related to sustainability (Oe et al., 2022). Many cultural practices are rooted in sustainable resource management, particularly among indigenous communities that have historically maintained ecological balance through their traditions (Ssozi, 2012; Uganda National Commission for UNESCO, 2010). Communities often possess valuable insights regarding their environmental challenges, which can inform and guide SD efforts. This approach ensures that solutions are tailored to local needs, increasing their effectiveness and likelihood of adoption (Michelsen & Wells, 2017; UNESCO, 2017). In this case, educational programmes should be tailored to reflect local cultural contexts, ensuring relevance to students' realities. Cultural heritage is also leveraged as an economic asset, generating income through tourism and related activities, thus contributing to local economies while promoting sustainability (World Bank, 2018).

An interdisciplinary approach is another ESD perspective in East Africa. An interdisciplinary approach to ESD is essential for equipping learners with the skills necessary to navigate and address the complexities of sustainability challenges. By integrating diverse disciplines, fostering collaborative learning environments, and leveraging technology, educational institutions can enhance their contributions to SDGs. This approach is vital in addressing complex sustainability challenges that cannot be effectively tackled through single-discipline perspectives (Berg & Verster, 2023). Interdisciplinary research allows students and faculty to explore sustainability issues from multiple angles, integrating knowledge from fields such as environmental science, economics, sociology, and technology (Mokski et al., 2023). This comprehensive approach helps in understanding the interconnectedness of various sustainability challenges and promotes critical thinking skills necessary for problem-solving in real-world contexts (Uganda National Commission for UNESCO, 2010). By encouraging collaboration among different academic departments, interdisciplinary research initiatives create a culture of shared knowledge, learning, and innovation, which are essential for advancing sustainable practices in various sectors (Araneo, 2024). For instance, the Virtual and Interactive Microgrids Learning Environment (VIMLE)—presented in Chapter 5.3—is a novel educational platform designed to enhance knowledge transfer in microgrid technologies in a sustainable context (Bogere et al., 2023). This collaboration not only enhances the educational experience for students but also leads to the development of more effective sustainability solutions that consider various stakeholder perspectives (Mokski et al., 2023). Interdisciplinary research often involves partnerships with industry, government, and community organizations. These collaborations enrich the educational experience and facilitate the dissemination of research findings into practice, thus enhancing the impact of ESD initiatives beyond academia (Araneo, 2024). Engaging in interdisciplinary research also helps students to develop essential skills such as teamwork, communication, and adaptability. These competencies are vital for future leaders who will navigate the complexities of SD in various professional contexts (Berg & Verster, 2023). As such, encouraging collaboration between departments promotes a culture of shared knowledge and innovation.

Furthermore, capacity-building is one of the perspectives of ESD in East Africa. Capacity-building for ESD is a multifaceted approach that involves various stakeholders working collaboratively to integrate sustainable practices into education systems (Ferguson et al., 2022). Through targeted programmes like those from UNECE and UNESCO, significant strides are being made towards achieving the SDGs and fostering a culture of sustainability in education worldwide (Teachers initiative in Sub-Saharan Africa (TISSA), 2014). In particular, through the A:RT-D Grids project, various students have been trained and supported to complete their doctoral studies. In this case, building the capacity of educators is vital for successful ESD implementation. Training programmes that focus on pedagogical strategies for teaching sustainability like VIMLE can, therefore, empower teachers to effectively engage students with diverse perspectives on SD. In the case of Uganda, ongoing professional development initiatives strive to equip educators with the necessary skills to integrate ESD into their teaching practices (Uganda National Commission for UNESCO, 2010). This initiative is aligned with global frameworks such as the United Nations SDGs, particularly Target 4, which emphasizes the importance of education in promoting SD.

5.1.4 Conclusion

In summary, ESD is essential for fostering a culture of sustainability in East Africa. By addressing existing challenges and implementing strategic initiatives, the East African region can enhance educational outcomes that contribute not only to individual development but also to broader societal progress towards sustainability. The perspectives on ESD in East Africa highlight the need for contextual relevance, interdisciplinary approaches, multi-stakeholder engagement, national frameworks, community practices, capacity-building initiatives, and global-local cooperation and integration. By embracing these perspectives, educators can foster a more sustainable future for their communities while equipping students with the knowledge, skills, values, and attitudes necessary to address complex sustainability challenges.

5.1.5 References

Agoi, R., & Nakato, F. (2012). *Project Progress Report. 2012*, 1–12.

Annan-diab, F., & Molinari, C. (2017). Interdisciplinarity: practical approach to advancing education for sustainability and for the Sustainable Development Goals. *International Journal of Management Education, 15*(2), 73–83.

Araneo, P. (2024). Exploring education for sustainable development (ESD) course content in higher education: a multiple case study including what students say they like. *Environmental Education Research, 30*(4), 631–660. https://doi.org/10.1080/13504622.2023.2280438

Berg, C. van den, & Verster, B. (2023). Integrating education for sustainable development into a sustainable-smart transdisciplinary learning framework. *The Independent Journal of Teaching and Learning, 8*(2).

Bogere, P., Bode, H., & Temmen, K. (2023). Work-In-Progress: Development of a Virtual and Interactive Microgrids Learning Environment for Microgrids Sustainability—The Case of East Africa BT. In M. E. Auer, W. Pachatz, & T. Rüütmann (Eds), *Learning in the Age of Digital and Green Transition* (pp. 671–679). Springer.

Carton, A. M., & Tewfik, B. A. (2016). A new look at conflict management in work groups. *Organization Science, 27*(5), 1125–1141. https://doi.org/10.1287/orsc.2016.1085

EAC Secretariat. (2022). *The East African Regional Bioeconomy Strategy 2021/22—2031/32.* 1–23.

Ferguson, T., Roofe, C., Cook, L. D., Bramwell-Lalor, S., & Hordatt Gentles, C. (2022). Education for Sustainable Development (ESD) infusion into curricula: influences on students' understandings of Sustainable Development and ESD. *Brock Education Journal, 31*(2), 63–84. https://doi.org/10.26522/brocked.v31i2.915

Green, B., & Green, B. (2016). *Community as a Context for Learning.* Northwest Center for Sustainable Resources. https://ate.is/downlands/122/Community%20as%20Context%20for%20Learning.pdf

Hargreaves, L. G. (2008). The whole-school approach to education for sustainable development: from pilot projects to systemic change. *Policy & Practice: A Development Education Review, 25290*, 69–74.

Michelsen, G., & Wells, P. J. (2017). *A Decade of Progress on Education for Sustainable Development.* UNESCO.

Mochizuki, Y., & Fadeeva, Z. (2010). Competences for sustainable development and sustainability: significance and challenges for ESD. *International Journal of Sustainability in Higher Education, 11*(4), 391–403. https://doi.org/10.1108/14676371011077603

Mokski, E., Filho, W. L., Sehnem, S., & Baltazar, J. (2023). Education for sustainable development in higher education institutions: an approach for effective interdisciplinarity. *Sustainability in Higher Education, 24*, 96–117.

Nantale, M., Atukunda, G., & Kyayemagye, F. (2022). Science teachers' remuneration and their job performance among government aided secondary schools in Mbarara, Uganda. *East African Journal of Management and Business Studies, 2*(1), 7–13. https://doi.org/10.46606/eajmbs2022v02i01.0005

Oe, H., Yamaoka, Y., & Ochiai, H. (2022). A qualitative assessment of community learning initiatives for environmental awareness and behaviour change: applying UNESCO Education for Sustainable Development (ESD) framework. *International Journal of Environmental Research and Public Health, 19*(6). https://doi.org/10.3390/ijerph19063528

Omisore, B. O., & Abiodun, A. R. (2014). Organizational conflicts: causes, effects and remedies. *International Journal of Academic Research in Economics and Management Sciences, 3*(6). https://doi.org/10.6007/ijarems/v3-i6/1351

Ssozi, L. (2012). Embedding education for sustainability in the school urriculum: the contribution of faith based organisation to curriculum development. *Journal of Sustainability Education, 3*(March).

Teachers' Initiative in Sub-Saharan Africa (TISSA). (2014). *Teacher Issues in Uganda: A Shared Vision for an Effective Teachers Policy* (September). UNESCO – IIEP Pôle de.

The East African Community. (2015). *East African Community Vision 2050: Regional Vision for Socio-economic Transformation and Development*. EAC IRC Repository, East African Community's Institutional Repository.

The Higher Education Academy Group. (2014). Education for Sustainable Development. *Encyclopedia of Quality of Life and Well-Being Research, June*, 1–31. https://doi.org/10.1007/978-94-007-0753-5_839

The Republic of Uganda. (2017). *The Republic of Uganda Decent Work Country Programme*. Kampala, Uganda, October 2012, 1–75.

Tikly, L. (2019). Education for Sustainable Development in Africa: a critique of regional agendas. *Asia Pacific Education Review, 20*(2), 223–237. https://doi.org/10.1007/s12564-019-09600-5

Uganda National Commission for UNESCO. (2010). *Uganda: Education for Sustainable Development Implementation Strategy Education for Sustainable Development Uganda Implementation Strategy*. UNESCO. 1–58.

UNECA. (2021). *Investing in Eastern Africa*. United Nations Economic Commission for Africa.

UNESCO, & UNEVOC. (2010). *Development in Technical and Vocational Education and Training in Selected Countries*. UNESCO & UNEVOC.

United Nations. (2024). *Investing in Eastern Africa: Opportunities and Challenges*. United Nations Economic Commission for Africa.

Uppsala University. (2017). *Visby Recommendations for Enhancing ESD in Teacher Education: UNESCO Global Action Programme on Education for Sustainable Development*. Swedish International Centre of Education for Sustainable Development (SWEDESD).

World Bank. (2018). GUIDANCE NOTE FOR BORROWERS: Environmental & Social Framework for IPF Operations. *World Bank*, 16. https:thedocs.worldbank.org/en/doc/87721522762050108-0290022018/original/ESFFramework.pdf

Zazu, C. (2016). Heritage education in post-colonial southern Africa: why it should be re-positioned within the framework of education for sustainable development (ESD). *Applied Microbiology and Biotechnology, 85*(1), 6.

5.2 The Use of Demonstrator Grids in Teaching Microgrid Development

Paul Bogere, Henrik Bode, Katrin Temmen

5.2.1 Introduction

Educational sites, designed to facilitate learning through direct practical experience and demonstrative objects, play a pivotal role in effective learning strategies. Such locations enable learners to directly connect theoretical knowledge with concrete, real-world experiences, thus promoting deeper understanding and lasting retention of the content taught (Kolb & Kolb, 2005). Authentic learning experiences, where learners can observe and experiment with real systems or processes, have been shown to enhance intrinsic motivation and long-term competence development (Nachtigall et al., 2022). Particularly in science, technology, engineering, and mathematics (STEM) education, interactive and practice-oriented learning environments are indispensable for comprehending complex concepts (Freeman et al., 2014).

Integrating ESD into curricula and community engagement requires more than just well-intended efforts from educators in classrooms; learners must also recognize its relevance. Experiencing the real benefits or bringing previously learned theory alive seems to be a promising way of increasing the awareness of the need of ESD to many different target groups including students, technicians, and the general community population.

Community learning centres with the possibilities of experiencing real microgrids have proven to be valuable spaces for fostering discussions on sustainability issues like climate change, resource management, and energy access (Araneo, 2024; Green & Green, 2016). Additionally, capacity-building on the non-academic level plays a crucial role in ESD, requiring collaboration among stakeholders to integrate sustainability into education systems (Ferguson et al., 2022). Demonstrator grids not only enhance technical skills in energy systems design and operation but also foster ownership and responsibility for sustainable energy solutions, supporting decentralized and resilient power systems in Eastern Africa (Kabasa et al., 2018, p.

14). But at the academic level, too, hands-on practice is highly demanded by future employers, as many of them identified a lack of competencies in this area (Kabasa et al., 2018, p. 12).

5.2.2 The Lwak Microgrid as a Local Multiplicator

A major aspect of the A:RT-D Grids project was the implementation of a real microgrid for research and demonstration purposes. Technical aspects of the Lwak microgrid are described in detail in Chapter 3.2.5.

The microgrid system serves as a powerful live demonstrator with a strong multiplier effect, providing educational, technical, and social benefits to various target groups. Its impact extends beyond the immediate operational goals and includes the promotion of knowledge transfer, community empowerment, and long-term sustainability.

Postulant Nuns: The system acts as a live laboratory for postulant nuns, enabling renewable energy, energy efficiency, and sustainability-related learning. By engaging with components like PV panels and inverters, the postulants acquire technical skills and energy literacy for future leadership roles.

Secondary Schools: The Lwak microgrid demonstrator enhances STEM education by providing hands-on experience in renewable energy-related technologies. Students learn about energy production, storage, and management hence inspiring interests in green energy careers and fostering data-driven decision-making skills.

Community Population: For the community population, the system acts as a demonstrator showing how efficient and clean renewable energy can be used. The fact that even the highest local religious authorities support this form of energy increases this effect and might lead local consumers or business owners to consider using a PV system.

Cooperation with SMA Sunbelt: Through its cooperation with SMA Sunbelt, the grid can be used for training local electricians or installers and can help in building local capacity. This is an important step for a sustainable clean energy system development.

Universities: Since the microgrid demonstrator officially belongs to the Ecolog Institute for social-ecological research and education, field trips for university students can be arranged to enable knowledge acquisition for sustainable energy.

5.2.3 Don Bosco Grid

As part of the graduate training (GT) in Dar es Salaam in 2023, the organizing team contacted the head of the Don Bosco Oysterbay Vocational Training Centre (VTC), a mission-driven non-profit organization that offers several courses to enable young people, often from disadvantaged backgrounds, to acquire real life skills. Some of the courses focus on electrical installations, with a very different approach of knowledge transfer compared to the universities involved in the project. Since money is a big issue in non-profit organizations, many materials need to have a dual use in generating or saving money. Most of the visual educational materials are sponsored posters from companies promoting their products. The same goes for their internal microgrid. Although there is potential to experience the whole energy grid up close, its clear prioritization to function as a cost-saving system over its demonstration purposes is evident in its missing several opportunities to increase the visibility and accessibility of components, even though only a very limited amount of money would be required for those.

In addition to the existing real microgrid at Don Bosco Oysterbay VTC, there exist various components that instructors may use to impart microgrid-related knowledge to trainees. Some of these components were used during the 2023 GT as revealed in Fig. 53. GT participants, in groups, were tasked with modelling a microgrid demonstrator.

Components used by the GT participants to model microgrid demonstrators include 12 V/30 W PV modules, 12/24 V 5 A charge controllers, 12 V/60 Ah (DoD 30 %) batteries, and 12 | 250 inverters. Other accessories used include 2 pole 63 A 240 V 50 Hz circuit breakers, 220–250 Volts LED AC lamps, 12 Volts LED DC lamp, lamp holders, single plastic MK boxes, single socket outlets, G-type top plugs, and 1 gang 1-way switches among others.

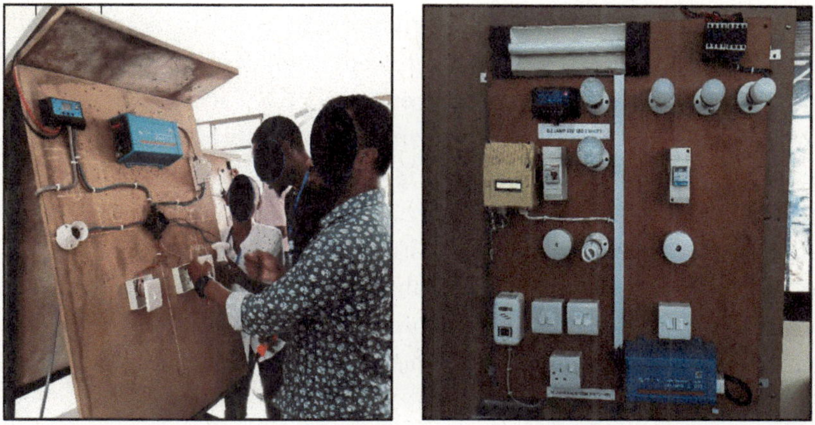

Fig. 53: Appliances and components for microgrid-related training at Don Bosco Oysterbay VTC in Dar es Salaam, Tanzania (reproduced from Bogere, 2024, p. 89). Left: GT participants tasked with modelling a microgrid demonstrator. Right: Assorted appliances for microgrids training.

Components used by the GT participants to model microgrid demonstrators include 12 V/30 W PV modules, 12/24 V 5 A charge controllers, 12 V/60 Ah (DoD 30 %) batteries, and 12 | 250 inverters. Other accessories used include 2 pole 63 A 240 V 50 Hz circuit breakers, 220–250 Volts LED AC lamps, 12 Volts LED DC lamp, lamp holders, single plastic MK boxes, single socket outlets, G-type top plugs, and 1 gang 1-way switches among others.

Relatedly, with the introduction of competency-based education in lower secondary curricula in EA countries, the A:RT-D Grids project developed microgrids course content for secondary school learners . During the implementation of the course at St. Mary's Lwak Girls' High School in Kenya, used similar components in one of the lessons as revealed in Fig. 54 and tasked learners with wiring a solar home system (SHS) to power a classroom TV. The wired SHS is comparable to a microgrid demonstrator given that it can be used to demonstrate and explain the operation of a microgrid.

The use of demonstrators is instrumental in enhancing learners' skills acquisition. As our examples reveal, the provision of laboratories through demonstration improves academic performance, and it is a recommended technique in teaching-learning environments with low apparatus and learning materials availability. Noting that most academic institutions in the East African region are characterized by inadequate facilities, demonstrators are therefore vital in boosting learners' interests and content engagement. Indeed, during the evaluation of the lesson summarized in Fig. 54, learners indicated that demonstrations »made the lesson fun, easy to understand and enjoyable« and that the tasks were interesting .

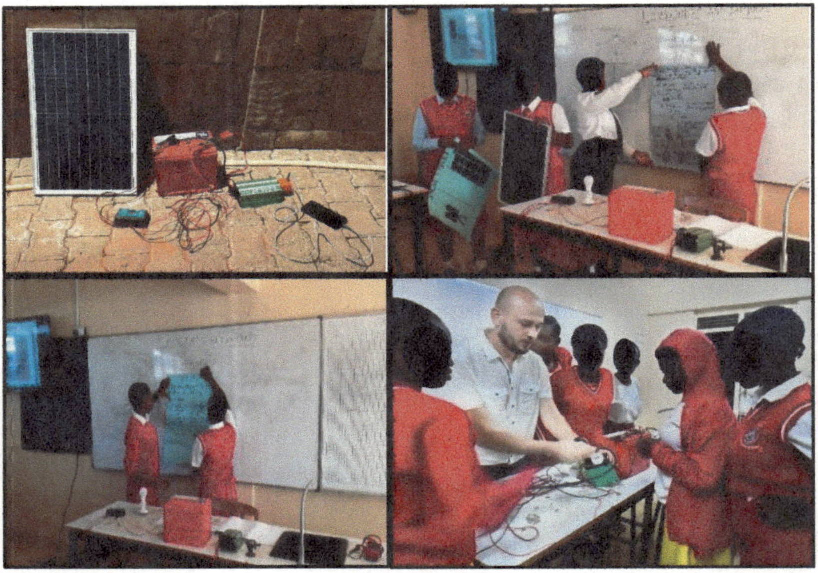

Fig. 54: A pictorial summary of the microgrids course lesson during which learners were tasked with wiring a microgrid demonstrator (photos reproduced from Belov, 2023, pp. 47–64).

5.2.4 Microgrid Demonstrators at German Universities

The importance of smaller energy grids is not only known in Eastern Africa. Knowledge transfer about those has been part of the curricula of German universities for a long time. At Paderborn University, for example, the number of real demonstrators has increased. While having only one grid demonstrator for teaching purposes in laboratory courses for about 30 years, and without any innovations, the situation has now changed considerably. Three additional microgrids operating at different levels and operated by different chairs for a large variety of purposes show the various ways in which demonstrators can be used. While the chair of power electronics uses a fully flexible research prototype in the range of several megawatts of power to investigate the control of energy grids in the age of distributed generation, the other freshly created grids have a clear demonstration and education focus. A small-scale microgrid with remote access was introduced at the chair of didactics of technology with the idea of enabling people around the world without access to a real grid to experience the advantages of one remotely. The last grid introduced by a fourth chair is designed to work, on the one hand, as a practical hands-on model to support lectures, and, on the other, as a demonstration grid for industrial partners and visitors to the university. The number of demonstrators that are currently built at Paderborn University highlights the importance of hands-on materials, even in times when AR/VR and simulations have become more and more a part of our everyday life.

5.2.5 References

Belov, A. (2023). Development, Implementation and Evaluation of Microgrids Course Materials for Secondary School Learners. Master's Thesis, Paderborn University.

Bogere, P. (2024). Knowledge Transfer for Microgrids Sustainability: The Case of East Africa. Doctoral Thesis, Paderborn University. https://doi.org/10.17619/UNIPB/1-2167

Bogere, P., Belov, A., & Temmen, K. (2024). Teaching and learning of microgrids for secondary school learners—a renewable energy sustainability strategy for East African communities. In E. de la P. Plaza, A. Blázquez-Soriano, R. Wang, & A. Sereni (Eds.), First International Conference on Sustainable Energy Education (SEED 2024) (pp.

128–136). Editorial Universitat Politècnica de València. https://doi.org/10.4995/SEED2024.2024.19007

Freeman, S., Eddy, S. L., McDonough, M., Smith, M. K., Okoroafor, N., Jordt, H., & Wenderoth, M. P. (2014). Active learning increases student performance in science, engineering, and mathematics. Proceedings of the National Academy of Sciences of the United States of America, 111(23), 8410–8415. https://doi.org/10.1073/pnas.1319030111

Kabasa, J. D., Asuman, S., Bukirwa, J., & Kisakye, H. (2018). Skills for Green Jobs in Uganda. International Labour Organization. https://www.ilo.org/sites/default/files/wcmsp5/groups/public/@ed_emp/@ifp_skills/documents/publication/wcms_706954.pdf

Kaneza, P., Nkurunziza, J. B., & Twagilimana, I. (2024). Effect of hands-on laboratory experiment and demonstration teaching techniques on Rwandan secondary school students' academic performance in solutions and titration. African Journal of Research in Mathematics, Science and Technology Education, 28(1), 44–56. https://doi.org/10.1080/18117295.2023.2275858

Kolb, A. Y., & Kolb, D. A. (2005). Learning styles and learning spaces: a review of multidisciplinary application of experiential learning theory in higher education. In R. R. Sims & S. J. Sims (Eds.), Learning Styles and Learning: A Key to Meeting the Accountability Demands in Education (pp. 45–92). Nova Science Publishers.

Nachtigall, V., Shaffer, D. W., & Rummel, N. (2022). Stirring a secret sauce: a literature review on the conditions and effects of authentic learning. Educational Psychology Review, advance online publication. https://doi.org/10.1007/s10648-022-09676-3

5.3 Developing a Virtual and Interactive Microgrids Learning Environment

Paul Bogere, Katrin Temmen

The evolution of virtual and interactive learning environments has been driven by the disruptions to education caused by the COVID-19 pandemic. While distance learning had existed before, the pandemic pushed academic institutions to maximize the use of electronic learning platforms (ELPs). These platforms are commonly used to share course materials and administer assessments, thus functioning as basic learning management systems (LMSs), characterized by the absence of real-time instructor-trainee interaction. To address this, instructors turn to platforms like Zoom and Google Meet though challenges remain in effectively demonstrating complex concepts online. In didactics of technology, illustrating concepts, laws, and theories helps to simplify otherwise abstract material and enhances learners' understanding and skills acquisition. Virtual and interactive learning environments are therefore seen as a potential solution. As Vergara et al. (2020) confirm, virtual and interactive learning environments enhance comprehension.

A Virtual and Interactive Microgrids Learning Environment (VIMLE) conceptual model, therefore, has been developed, as detailed in subsection 5.3.3. VIMLE is a didactically designed online system for hosting and enabling access to developed content and software-based virtual laboratories in addition to facilitating instructor–trainee interaction specifically for transferring and acquiring microgrids-related knowledge, skills, and competencies. The role of VIMLE in ESD and its envisaged contribution to SDGs are presented in subsection 5.3.1. Relatedly, the use of mobile phones and text-based tools (Grönlund & Islam, 2010) coupled with the teaching and learning of residential electrical installations outside of formal educational settings (Araújo, Guimarães, et al., 2024; Araújo, Sales, et al., 2024) reveal that VIMLE can benefit not only traditional students but also other groups, such as electrical energy consumers. Thus, VIMLE can be applicable in both formal and informal educational contexts. Stakeholders that can be trained using VIMLE are therefore described in 5.3.2. VIMLE implementation challenges and opportunities are explained in 5.3.4 and the conclusion is presented in 5.3.5.

5.3.1 The Role of VIMLE in Contributing to Sustainability

VIMLE contributes to sustainability through its envisaged role in ESD and the realization of selected SDGs. ELP and LMS, such as VIMLE, are crucial in ESD. They provide accessible environments for knowledge transfer, sharing, collaboration, and skill development (Donath et al., 2020). The platforms, therefore, support learning by enabling access to digital resources and multimedia content and facilitate gamification techniques to enhance instructor–trainee engagement (Donath et al., 2020). Gamification techniques involve technologies that offer interactive didactic instruction. Such technologies inlcude augmented reality (AR) as well as virtual reality (VR) and are particularly effective for sustainability topics, promoting action-oriented understanding and knowledge transfer (Ebinger et al., 2022). ELP and LMS also facilitate essential ESD skills development. Such ESD skills include critical thinking, systemic thinking, partnership-building, and participatory decision-making (Zhang et al., 2020). With ESD skills acquired, the Human Resource (HR) is developed. HR is vital in achieving sustainability and economic development in communities (Zhang et al., 2020, p. 3). ESD is therefore vital in achieving SDGs. Indeed, Leal Filho (2021) and Zhang et al. (2020) emphasize that learning is an essential tool in achieving SD and thus SDGs.

With the development of VIMLE, knowledge, skills, and competencies shall be acquired by the users, trainees in particular, hence contributing to HR development. VIMLE is developed to enhance microgrids' sustainability by facilitating knowledge transfer related to microgrids' operation and application. Such transfer of knowledge plays a crucial role in building local HR capacity, ensuring two key outcomes: (1) the development of a skilled pool of future energy professionals capable of designing, installing, managing, and maintaining microgrid systems; and (2) raising awareness and equipping community members with the skills needed to adopt new perspectives regarding the potential of microgrids. Rather than restricting their use to basic services such as lighting, phone charging, or radio operation, communities can be empowered to create innovative solutions and enterprises that respond to local needs. This approach encourages productive uses of energy (PUE), thereby supporting more efficient utilization of electricity, stimulating demand, and increasing both consumption and household incomes, as energy access is linked to income-generating activities. As highlighted by Booth et al. (2018), PUE is a critical success factor for microgrids. Likewise, Boche et al. (2022, pp. 4, 17–18) emphasize that knowledge dissemination is central to ensuring microgrid sustainability. By adopting a didactic approach, focused on structured instruction and training, VIMLE can equip trainees with essential skills and competencies. This educational strategy aligns with SDG 4, specifically target 4.7, while simultaneously supporting the achievement of SDG 7 by advancing access to sustainable energy.

5.3.2 VIMLE as a Learning Tool—Who are the Stakeholders?

As mentioned in 5.3.1, VIMLE aims to contribute to microgrids' sustainability through knowledge transfer. Microgrids, as envisaged solutions to improve electrical energy access, can better achieve the purpose if installed in rural communities far from the centralized grid—mostly in developing countries. It is therefore instructive that rural communities are engaged in microgrids-related training. To this end, consumers of electrical energy services in rural communities are the primary stakeholders. Consumers therefore need to be trained to acquire basic microgrids knowledge and be empowered to make informed decisions regarding the use of electrical energy to suit their requirements and contribute to microgrids' sustainability. To achieve this, consumers in rural communities were engaged to understand their local challenges, training interests, training needs, and transferrable skills. Consequently, consumers need to acquire relevant knowledge and skills, perform simple electrical installations, and acquire entrepreneurship knowledge and skills for income generation, among other things. It is therefore instructive that VIMLE enables consumers to acquire skills in line with their challenges, interests, and needs. VIMLE, as a learning tool, should enable consumers to: identify components of a microgrid system; carry out simple maintenance tasks of selected microgrid components; identify necessary electrical installation accessories; perform basic household electrical installation; apply fault-finding and rectification procedures in electrical installations; discuss local challenges that are potential entrepreneurial opportunities; and set up income-generating businesses that respond to local challenges.

As a microgrids knowledge transfer, and thus a microgrids' sustainability strategy, secondary school learners aged 14 to 18 need to be trained as well (cf. Chen et al., 2014). As such, secondary school learners of 14 to 18 years of age make up the second category of stakeholders. It is envisioned that on the acquisition of microgrids-related knowledge and skills, secondary school learners can apply the same in their respective communities. Based on the secondary schools' physics syllabi from Kenya, Tanzania, and Uganda, appropriate microgrids-related knowledge, skills, and competencies for secondary school learners were crafted. As a learning tool, VIMLE should therefore enable secondary school learners to be able to: design solar home systems; recommend suitable and efficient household electrical accessories and appliances; and perform basic domestic electrical installations and troubleshooting.

To ensure the availability of energy engineers to design, install, operate, and maintain microgrids in East Africa (EA), students and graduates of the Bachelor of Science in Electrical Engineering (BSc Elec Eng) need to be trained. Hence, the third category of stakeholders. If BSc Elec Eng students are skilled, then locally trained HR to implement even complex microgrid projects and ensure their sustainability is built. Based on BSc Elec Eng curricula, junior and senior students, lectures of power-related courses from selected universities in EA and microgrid operators' insights, content to support microgrids-related workplaces and provide the required employability skills for day-to-day tasks is developed. The content should enable BSc Elec Eng students and graduates to: design microgrid energy systems conscious of electrical energy requirements, and apply Artificial Intelligence (AI), Industrial Internet of Things (IIoT), and sensor technology to microgrids' operation and reliability optimization, among others. As a learning tool, therefore, VIMLE should enable BSc Elec Eng students and graduates to be able to: produce models of microgrid systems; design microgrid systems; acquire skills related to microgrids' installation, commissioning, and operation; design AI-based energy management solutions; develop and implement logic control algorithms to optimize microgrids' operation.

Therefore, based on the foregoing discussion, VIMLE trainee stakeholders include consumers of electrical energy services, secondary school learners aged 14 to 18, and BSc Elec Eng students and graduates. In developing the VIMLE content to guide stakeholder-specific skills acquisition, the following are consulted: consumers, lower secondary physics syllabi, BSc Elec Eng curricula, junior and senior students, lecturers of power courses, and microgrid operators. In essence, stakeholder contextual relevance in contributing to microgrids' sustainability is considered in crafting specific skills. To this end, VIMLE development is described in 5.3.3.

5.3.3 VIMLE Conceptual Model

The VIMLE conceptual model integrates the envisaged VIMLE attributes, the didactic foundation, and the technology amalgamation. The desirable attributes of VIMLE are synchronous, asynchronous, cooperative, and experiential learning. According to Bogere et al. (2023, p. 673) the VIMLE synchronous learning feature enables real-time interaction between instructors and trainees in a virtual classroom. In the same way, Francescucci and Rohani (2019) add that the classroom is instructor-led, takes place online, and that based on their study involving 698 students taught across two semesters, the virtual classroom pedagogical technique can achieve equivalent »student performance and engagement outcome« levels as traditional in-person instruction. The research findings in Francescucci and Rohani (2019, pp. 64–65) indicate no substantial statistical differences in the average performance levels of students who attended the physical and the virtual classes. Relatedly, a study involving 12 students conducted by Bilad (2023) for one semester indicates no considerable statistical difference in the overall performance between the online asynchronous trainee self-directed and the synchronous instructor-led, face-to-face learning approaches. Based on available studies, therefore, the asynchronous and synchronous attributes can enable VIMLE users to achieve comparable learning outcomes as trainees who attend physical classes. Another VIMLE desired attribute closely linked to asynchronous is experiential learning. Experiential learning can enable trainees to engage in simulations of real-world scenarios that relate to microgrids. For example, trainees can interact with 3D virtual microgrid component models of solar panel generators, and battery storage systems, among other main and auxiliary components of microgrid systems. This allows VIMLE trainees to experience operational aspects of microgrids in a controlled environment and as such it calls for the development of virtual VIMLE laboratories (V-Labs) to enable trainees to engage in practical assignments. Practical problem-solving activities, such as balancing supply and demand, troubleshooting microgrid faults, and case studies of microgrid and consumer electrical installations, are incorporated to help trainees apply their knowledge to realistic scenarios and develop hands-on skills. Another desired attribute of VIMLE is cooperative learning that enables peer-to-peer trainee collaboration. The cooperative attribute enables trainees to consult one another and execute assignments and group projects jointly, leading to enhanced learning. To enable the realization of these desired attributes, the VIMLE system should have multimedia (video, audio, and graphics) live-streaming capabilities to facilitate real-time instructor–trainee interaction. In addition, the system should enable: (1) instructors to share instructional materials; (2) communication between system users, for example, via a chat window that links instructors' and trainees' email addresses and other social media platforms.

The didactic foundation not only guides the trainees' interaction with VIMLE but also shapes how they interact with the training content while ensuring relevance and effective construction of the intended knowledge, skills, and competencies. The knowledge, skills, and competencies that trainees ought to acquire at the end of any VIMLE-supported learning activity are composed as Intended Learning Outcomes (ILOs). The composing of ILOs is guided by the revised Bloom's taxonomy (Krathwohl, 2002; Wilson, 2016). A constructive alignment (CA) concept is then applied to establish: (1) The electronic teaching-learning activities (e-tivities) that enable trainees to acquire envisaged competencies; (2) The assessment and feedback methods that enable the instructor to ascertain that trainees have acquired the intended competencies; (3) A linkage between ILOs, e-tivities, and the assessment and feedback methods. Some researchers (e.g., Thurner & Böttcher, 2021, p. 4) regard the CA concept as a modern didactic tool for knowledge constructivism. Applying the CA concept that appears in Bogere et al. (2023, p. 676), a snippet of the CA frame linking e-tivities, ILOs, and assessment methods is indicated in Fig. 56. The CA frame is vital in developing V-Labs and the associated e-tivities.

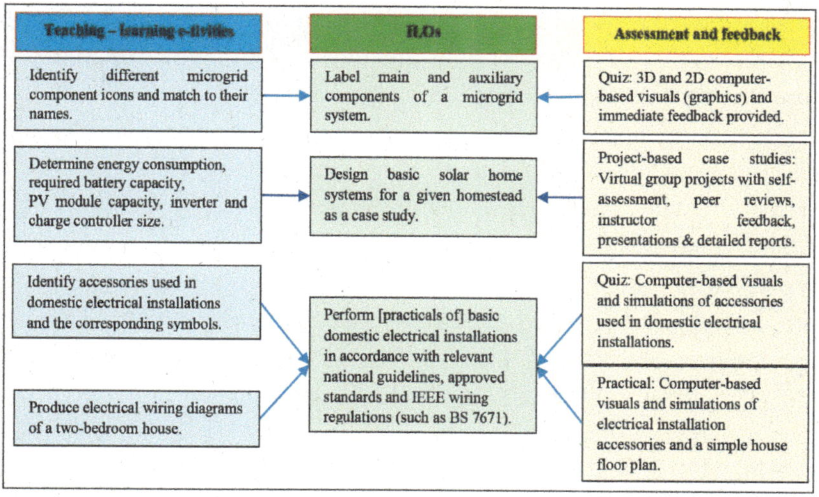

Fig. 55: A CA frame for selected ILOs (adapted from Bogere et al., 2023, p. 676).

V-Labs are vital in causing a deeper understanding of essential microgrid knowledge (Krajčovič et al., 2021). Virtual three-dimensional (3D) models are used for this purpose. The 3D models are developed using Unity 3D and Vuforia Augmented Reality (AR) Software Development Kit (SDK). The Visual Studio development environment is integrated into Unity and is vital during V-Labs development to enable interactivity. In addition, Virtual Reality (VR) that enables trainees to interact with virtual labs and the AR technology that enables virtual and real worlds to co-exist are employed in developing V-Labs. Developed V-Labs are then added to the V-Labs Management System (MS) of VIMLE's core system design.

5.3.3.1 VIMLE's Core System Design based on the MVC Architecture–

In addition to the V-Labs MS, other MSs form part of the VIMLE's core system design which is based on the Model View Controller (MVC) architecture (Huang & Chao, 2005; Pandit et al., 2010). The MVC architecture shows relationships between information, information display, and user command objects that trigger protocols to retrieve stored information (Reenskaug et al., 2001, p. 310). Fig. 56 (reproduced from Bogere, 2024, p. 118) illustrates a VIMLE's core system design based on MVC architecture. The system design consists of four levels including resource, VIMLE services, common service, and the presentation level. The system design describes how different internal VIMLE components work in enabling a system user to access the required information.

The resource level manages resources in the services level by enabling system users to retrieve information from the repository server through the VIMLE web server. The resources in the repository include the instructional materials and the V-Labs. The resource level enables access to the VIMLE system resources and is therefore responsible for the VIMLE system logic regarding access to the repository.

The VIMLE services level, on the other hand, contains all the VIMLE information management system (MS) components. The VIMLE services level combines related information, for ease of access and efficient operation of the VIMLE system, in MS components. Information related to all accounts for system users such as administrators, instructors, learners, and guests is arranged in the Accounts MS (AMS). The Content MS (CMS) contains all instructional materials including the developed content. Other MS components include the V-Labs MS, the Synchronous, Asynchronous, and Cooperative Classrooms MS (SACMS); the Tests, Exams, and Assignments MS (TEAMS); the Assessment, Results, and Feedback MS (ARFMS); the Remote Laboratories MS (R-Labs MS); Other Online Platforms MS (OOPMS) that provide microgrids-related training; and any other Information and Communication MS (ICMS). The VIMLE services level, therefore, delivers system application features such as registration, login, and content access permissions among others to VIMLE users based on users' needs and rights. This is achieved by utilizing the functionalities provided by the implemented VIMLE components.

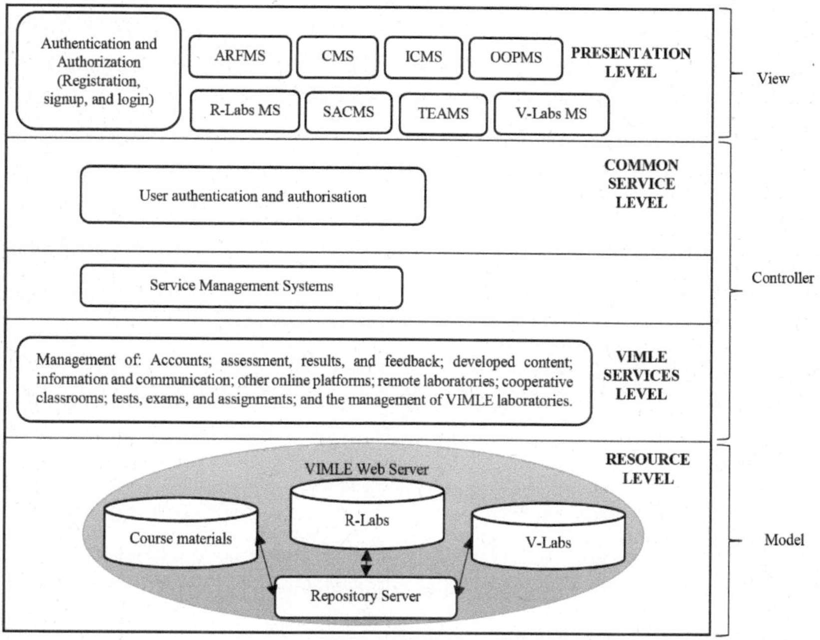

Fig. 56: VIMLE core system design based on the MVC architecture (reproduced from Bogere, 2024, p. 118; originally adapted from Pandit et al., 2010, p. 57).

Users' authentication and authorization is accomplished by the common service level. The common service level contains the service MS representing the VIMLE services level. The common services level queries the VIMLE services level and transmits the required information to users through the presentation level. The common service level therefore bridges the presentation and the VIMLE services levels. The common service and VIMLE services levels form the MVC architecture controller part. The controller part, then, serves as a link between the model and view within the

MVC architecture. Acting as an intermediary, the controller processes received user requests, relays them to the model, and returns responses either directly to the users or to a repository for storage.

The presentation level manages the interaction between users and the VIMLE services level. The interaction follows user authentication and authorization; it is facilitated by the common service level and realized through graphical user interfaces (GUIs). Users, therefore, engage with visual displays such as icons or tabs on their devices to access VIMLE services. On navigating the VIMLE platform's website, user requests and responses are transmitted as HTTP. This application-layer protocol handles the exchanges as hypermedia documents, for example, HTML. Consequently, the flow of information between users and the VIMLE platform occurs as HTML documents exchanged between the web browser and the VIMLE server. Whereas the VIMLE core system design is based on MVC architecture, the VIMLE architectural design, which provides a broader outlook of VIMLE system operation, is based on the WebRTC technology.

5.3.3.2 VIMLE Architectural Design based on WebRTC Technology

WebRTC is a web-based technology that allows web clients (devices of VIMLE platform users in this case) that support HTML5 to communicate, interact, or video conference in real time. Clients need only a browser that supports WebRTC technology. Examples of browsers that support WebRTC technology include Chrome, Firefox, and Safari, among others. A WebRTC architecture, with two clients and a centralized VIMLE server, is shown in Fig. 57.

Referring to Fig. 57, the server receives and responds to client service requests. Client A signals to the server indicating the need to interact with Client B. The server sends information of A to Client B who in turn sends signal information to the server. If Client B is willing to interact with Client A, the server forwards details of Client B to A. Upon receipt of Client B's particulars, Client A uses that information to set up a peer-to-peer connection between A and B. This is based on the Session Initialization Protocol (SIP) (Edan et al., 2017, p. 77).

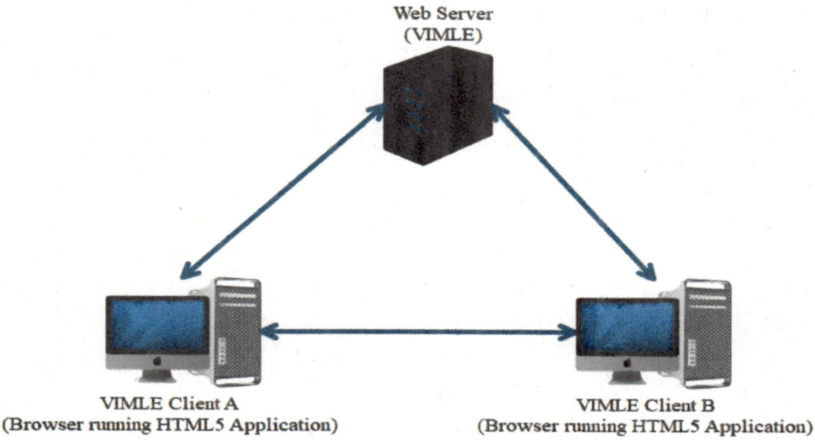

Fig. 57: Illustration of the VIMLE architectural design based on WebRTC technology (reproduced from Bogere, 2024, p. 116; Bogere et al., 2023, p. 674).

Based on the foregoing discussion and considering more than two clients, local clients, and remote clients on different IP networks, the proposed VIMLE architectural design based on WebRTC technology is as demonstrated in Fig. 58. The architecture indicates the developed instructional course materials in CMS, the VIMLE laboratories in V-Labs MS, the remote laboratories in R-Labs MS and other MSs all accessible via the VIMLE web server based on WebRTC technology.

Architectural frameworks based on WebRTC technology have many advantages. For example, they can run on any device. This is because their APIs contain the majority of browsers, so the system (or app) can run on any device with any of the major browsers installed.

However, although WebRTC uses end-to-end encryption, one notable disadvantage is that WebRTC leaks once connected to a VPN. This problem can be solved by ensuring that once connected to a VPN, the VPN extension is installed on the browser of choice and that the leak protection functionality is enabled. This justifies the inclusion of a network-based firewall in Fig. 58 to ensure access authorization only to clients who meet a prescribed set of security guidelines. One of such guidelines could

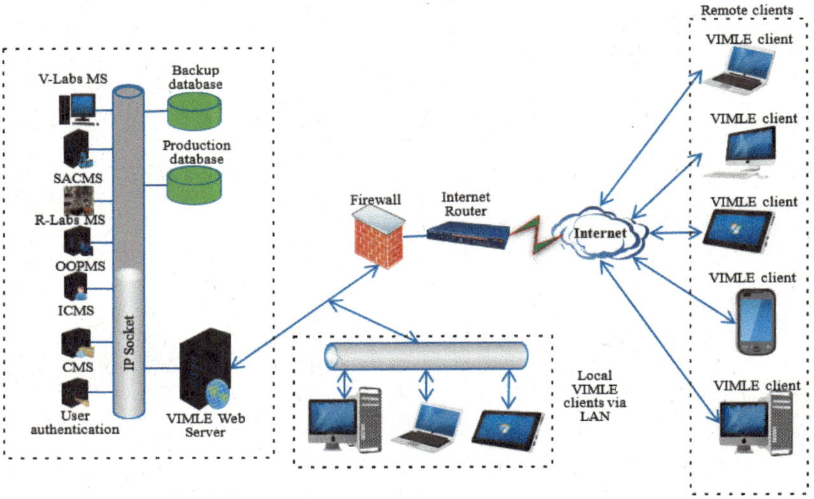

Fig. 58: VIMLE architectural design illustration indicating selected VIMLE service MSs, local and remote clients (reproduced from Bogere, 2024, p. 120).

be limiting clients to a particular VPN. Once a client is logged in and the leak protection functionality is enabled, the communication is secure.

To put the foregoing discussion into perspective, a VIMLE system layout is presented in Fig. 59. The layout illustrates both the back and front ends. The back end shows the resource and service levels of VIMLE's core system design. The HTTP/Websockets that underpin WebRTC technology operation are revealed in the system layout's back end. Similarly, the front end shows the presentation level of VIMLE's core system. The presentation level depicts the VIMLE platform impression from the users' point of view. The front end, therefore, reveals the interface tabs that VIMLE platform end-users can interact with while using the system. The course materials tab, for example, enables access to the stakeholder-specific developed content while the V-Labs tab enables access to the virtual VIMLE laboratories. From the designer's point of view, the front end illustrates the users and the presentation level of VIMLE's core system design. Decisions about presentation logic are made in the presentation level in the view part of the MVC. The decisions include the position, colour, and size of text, among others.

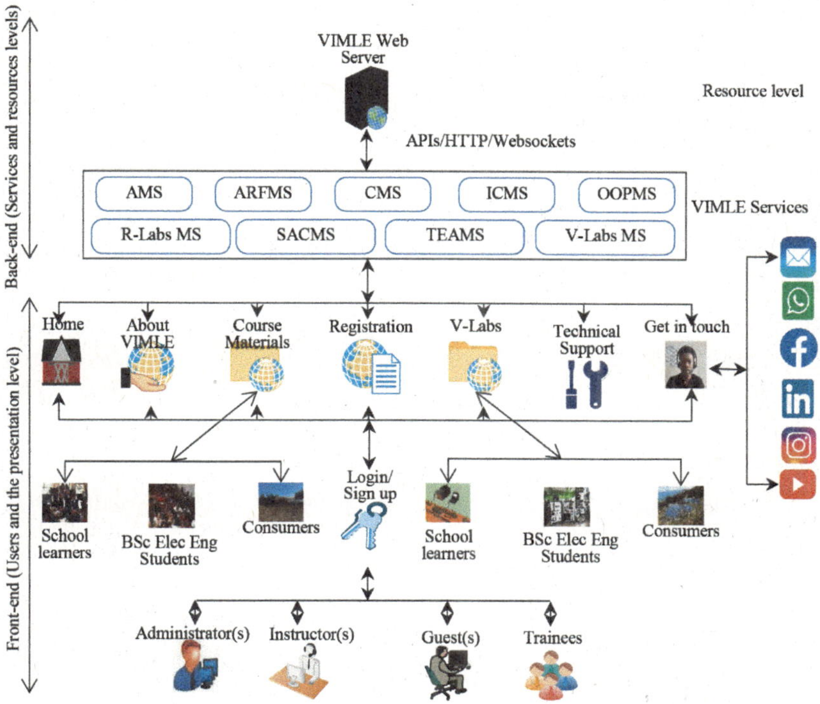

Fig. 59: VIMLE system layout (reproduced from Bogere, 2024, p. 122; originally adapted from Bogere et al., 2023, p. 677).

5.3.4 VIMLE Implementation Challenges and Opportunities

At inception, the envisaged stages of VIMLE implementation included the conceptual design, modelling of 3D microgrid components and V-Labs, deployment, testing, and evaluation. The conceptual design is complete and partial results are presented in 5.3.3. The second vital stage of 3D and V-Labs development suffered a technical expertise challenge. The plan was to employ graduate research assistants to write core computer algorithms to develop 3D models and V-Labs. The hired research assistants »learnt on the job«; they focused more on their Master's studies and the stage took longer than anticipated. The other alternative is to procure the services of computer programmers to develop professional 3D models and V-Labs, but these are costly. Therefore, financial constraints and sustainability after the A:RT-D Grids project are other challenges that threaten VIMLE realization.

Despite the indicated challenges, VIMLE is considered a feasible tool for microgrids-related knowledge transfer, skills acquisition, and hence local HR development in the EA region. VIMLE can certainly reduce the microgrids-related knowledge gap between urban and rural populations. However, for electrical energy consumers in rural communities of developing countries, infrastructural constraints and technological barriers that result in limited internet access and power supply, in addition to the digital divide, hinder the realization of the knowledge gap reduction. In situations of disasters or pandemics like COVID-19, ELPs such as VIMLE are innovative training platforms and can enable continuity of skills acquisition.

5.3.5 Conclusion

In summary, this subchapter presents the development of VIMLE and reveals VIMLE's contribution to sustainability through advancing ESD and SDGs. By providing a virtual and interactive platform for microgrids-related knowledge transfer and thus enabling acquisition of skills, VIMLE directly contributes to SDG 4. By promoting knowledge transfer, VIMLE contributes to local skilled HR development. Consequently, the skilled HR contributes to sustainable microgrids thereby improving electrical energy access levels, hence contributing to SDG 7. The VIMLE platform serves a broad range of stakeholders, including BSc Elec Eng students, secondary school learners, and electrical energy consumers. These stakeholders, as VIMLE users, can benefit from a deeper understanding of microgrid technology, acquire requisite skills, and hence be empowered to contribute to microgrids' sustainability and innovation in their respective communities.

This subchapter further highlights the VIMLE conceptual model, which encompasses its core attributes such as synchronous, asynchronous, cooperative, and experiential learning. It is worth noting that, based on available studies, asynchronous and synchronous attributes can enable VIMLE users to achieve comparable learning outcomes as trainees who attend physical classes. VIMLE enables collaborative learning and provides interactive modules. Grounded on the didactic foundation, VIMLE leverages technologies like VR and AR to create an engaging and effective learning experience supported by 3D models and V-Labs. The subchapter further describes the VIMLE core system design based on MVC architecture. The system design focuses on seamless integration of various learning tools and the protocol for user information accessibility and retrieval, while VIMLE architectural design, based on WebRTC, ensures flexibility and is concerned with the broader outlook of VIMLE system operation. It should be noted, though, that VIMLE implementation comes with challenges, such as inadequate technical expertise, financial and infrastructural constraints, and thus uncertainty in terms of sustainability. Despite these, VIMLE presents opportunities for enhancing sustainable education such as enabling microgrids' knowledge transfer, skills acquisition, local HR development, and hence fostering innovation in the energy sector even during crisis situations or pandemics like COVID-19.

5.3.6 References

Araújo, J. F., Guimarães, R. T., Silva, C. A., & Silva, H. D. (2024). Virtual environment for assisting learning of residential electrical installations. *IEEE Revista Iberoamericana de Tecnologias Del Aprendizaje, 19*, 45–50. https://doi.org/10.1109/RITA.2024.3368362

Araújo, J. F., Sales, L. M. R., Silva, C. A., & Silva, H. D. (2024). virtual environment for teaching and learning in residential electrical installations projects. *IEEE Revista Iberoamericana de Tecnologias Del Aprendizaje, 19*, 39–44. https://doi.org/10.1109/RITA.2024.3368368

Bilad, M. R. (2023). Enhancing engineering electromagnetics education: a comparative analysis of synchronous and asynchronous learning environments. *International Journal of Essential Competencies in Education, 2*(1), 66–74. https://doi.org/10.36312/ijece.v2i1.1369

Boche, A., Foucher, C., & Villa, L. F. L. (2022). Understanding microgrid sustainability: a systemic and comprehensive review. *Energies, 15*(8), 1–29. https://doi.org/10.3390/en15082906

Bogere, P. (2024). *Knowledge Transfer for Microgrids Sustainability: The Case of East Africa.* Doctoral Thesis, Paderborn University.

Bogere, P., Bode, H., & Temmen, K. (2023). Work-in-progress: development of a virtual and interactive microgrids learning environment for microgrids sustainability—the case of East Africa. In M. E. Auer, W. Pachatz, & T. Rüütmann (Eds.), *Learning in the Age of Digital and Green Transition. ICL 2022. Lecture Notes in Networks and Systems* (Vol. 1, pp. 671–679). Springer Nature.

Booth, S., Li, X., Baring-Gould, I., Kollanyi, D., Bharadwaj, A., & Weston, P. (2018). *Productive Use of Energy in African Microgrids: Technical and Business Considerations*. National Renewable Energy Laboratory. https://docs.nrel.gov/docs/fy18osti/71663.pdf

Chen, C. F., Tomsovic, K., & Aydeniz, M. (2014). Filling the pipeline: power system and energy curricula for middle and high school students through summer programs. *IEEE Transactions on Power Systems, 29*(4), 1874–1879. https://doi.org/10.1109/TPWRS.2013.2293752

Donath, L., Mircea, G., & Rozman, T. (2020). E-learning platforms as leverage for education for sustainable development. *European Journal of Sustainable Development, 9*(2), 1–19. https://doi.org/10.14207/ejsd.2020.v9n2p1

Ebinger, F., Buttke, L., & Kreimeier, J. (2022). Augmented and virtual reality technologies in education for sustainable development: an expert-based technology assessment. *Zeitschrift Fur Technikfolgenabschatzung in Theorie und Praxis / Journal for Technology Assessment in Theory and Practice, 31*(1), 28–34. https://doi.org/10.14512/tatup.31.1.28

Edan, N. M., Al-Sherbaz, A., & Turner, S. (2017). Design and evaluation of browser-to-browser video conferencing in WebRTC. *Global Information Infrastructure and Networking Symposium, GIIS 2017*, 75–78. https://doi.org/10.1109/GIIS.2017.8169813

Francescucci, A., & Rohani, L. (2019). Exclusively synchronous online (VIRI) learning: the impact on student performance and engagement outcomes. *Journal of Marketing Education, 41*(1), 60–69. https://doi.org/10.1177/0273475318818864

Grönlund, Å., & Islam, Y. M. (2010). A mobile e-learning environment for developing countries: the Bangladesh virtual interactive classroom. *Information Technology for Development, 16*(4), 244–259. https://doi.org/10.1080/02681101003746490

Huang, F. M., & Chao, M. (2005). An architecture of virtual environment for e-learning (AVEE). *5th IEEE International Conference on Advanced Learning Technologies, ICALT 2005*, 148–149. https://doi.org/10.1109/ICALT.2005.49

Krajčovič, M., Gabajová, G., Matys, M., Grznár, P., Dulina, Ľ., & Kohár, R. (2021). 3D interactive learning environment as a tool for knowledge transfer and retention. *Sustainability (Switzerland), 13*(14). https://doi.org/10.3390/su13147916

Krathwohl, D. R. (2002). A revision of Bloom's taxonomy: an overview. *Theory Into Practice, 41*(4), 212–219.

Leal Filho, W. (2021). Non-conventional learning on sustainable development: achieving the SDGs. *Environmental Sciences Europe, 33*(1). https://doi.org/10.1186/s12302-021-00525-8

Pandit, S., Acharya, H., Sanvedi, A., Nahar, P., Soni, S. K., Saxena, P., & Ramani, A. K. (2010). An architecture solution for e-learning system—ESSA. *International Conference on Technology for Education, T4E 2010*, 56–62. https://doi.org/10.1109/T4E.2010.5550061

Reenskaug, T., Wold, P., & Lehne, O. A. (2001). *Working with Objects: The OOram Software Engineering Method.* [Reprint ed.]. Addison-Wesley.

Thurner, V., & Böttcher, A. (2021). Designing an architecture for structuring didactic concepts, methods and tools. *ICL2021—24th International Conference on Interactive Collaborative Learning, September*, 864–875.

Vergara, D., Rubio, M. P., Lorenzo, M., & Rodríguez, S. (2020). On the importance of the design of virtual reality learning environments. In R. Gennari, P. Vittorini, F. De la Prieta, T. Di Mascio, M. Temperini, R. A. Silveira, & D. A. O. Carranza (Eds), *Methodologies and Intelligent Systems for Technology Enhanced Learning, 9th International Conference. MIST4TEL 2019. Advances in Intelligent Systems and Computing* (Vol. 1007, pp. 146–152). Springer. https://doi.org/10.1007/978-3-030-23990-9_18

Wilson, L. O. (2016). *Anderson and Krathwohl Bloom's Taxonomy Revised. Understanding the Revised Version of Bloom's Taxonomy* (1956). https://intranet.ecu.edu.au/__data/assets/pdf_file/0010/772867/Understanding-the-New-Blooms-Taxonomy-Wilson-2016.pdf

Zhang, T., Shaikh, Z. A., Yumashev, A. V., & Chład, M. (2020). Applied model of e-learning in the framework of Education for Sustainable Development. *Sustainability (Switzerland), 12*(16). https://doi.org/10.3390/SU12166420

5.4 Approaches to Community Learning in Microgrid Development

Teddy Mangeni, Paul Bogere, Henry Asiimwe, Christine Freitag

5.4.1 Introduction

Talking about community learning in microgrid development demands a look into basic concepts that frame this learning. As the focus is on a special setting—i.e., learning about, within, and for community life in a renewable energy setting—the first framework to rely on is the ESD framework as described in chapter 5.1. In order to be able to make ESD work, its contextuality has been highlighted for many good reasons. Linked to very different contexts, but combined with energy transformation, we can see another framework which may be called transformational learning (cf. Asiimwe et al. 2024, p. 247). As we see the chances and the challenges that occur with the implementation of community microgrids, it is clear that these implementations do and must appear as socio-technical transformations that need response. The process of finding answers and establishing solutions may be seen as the second side of transformational learning. It will (ideally) transform people in their attitudes and actions, where it might lead to more sustainability. However, as evident in chapter 4, different stakeholders react differently, and there are conflicts that show the process not always to be a smooth one. Failures, even if called technical, are often linked to human actors.

The following subchapters will help to sharpen the view on ESD community learning, as it is often overlooked in ESD discourse but is of particular interest in community energy development. Microgrid communities share transforming realities and challenges and need to negotiate power acceptance among themselves and with other stakeholders involved (Mangeni, 2024). Subchapter 5.4.2 gives a broad overview of existing methods and media of community learning. Subsection 5.4.3 considers communities as key resources for learning. Subchapter 5.4.4 explores the discourses that emerge in microgrid communities concerning the gendered character of access and strategies of learning.

5.4.2 Media and Methods of Community Learning in Microgrid Communities

Community learning plays an important role in microgrid development, particularly in East Africa, where decentralized microgrids are regarded as solutions to improve electrical energy access and propagate educational, economic, health, and social progress. The practice of training and skilling inhabitants in local communities ensures that electrical energy consumers can install, maintain, and optimize microgrid systems, in turn promoting sustainability and ownership. Various media and methods are employed in skills and knowledge transfer activities. These range from the traditional face-to-face training, on-the-job training, radio broadcasting, to online platforms, among others.

Traditional face-to-face training is a common didactic method employed by energy developers, training consultants, and community-led initiatives to achieve microgrids-related skills and knowledge transfer. Face-to-face trainings are normally organized in local community centres. To avert learning by rote and enhance knowledge retention, hands-on experiments, demonstrations, and excursions to live microgrids should be employed to supplement the face-to-face training method (cf. Bogere, Temmen, et al., 2023; Kaneza et al., 2024; Krüger et al., 2022). As revealed by Krüger et al., (2022, p. 190), experiments promote situational interests and support social learning. Further, open homes and local cooperatives can be leveraged to form electrical energy user groups and organized to establish knowledge-sharing forums. Following training by instructors including experienced technicians, peer-to-peer learning may take place where trained consumers may train fellow community members in their respective homes and forums. Such grassroots initiatives promote social cohesion and microgrids sustainability (cf. Heiskanen et al., 2017; KFW Development Bank et al., 2020, p. 60).

On-the-job training programmes play a significant role in imparting practical skills (KFW Development Bank et al., 2020, p. 66). International development organizations such as the German Society for International Cooperation (GIZ), Japan International Cooperation Agency (JICA), the United Nations Development Programme (UNDP), and Power Africa partner with local enterprises and microgrid energy developers to offer real-world exposure through apprenticeships (KFW Development Bank et al., 2020; Mwakisisya, 2020). This approach enables local community inhabitants to gain first-hand experience in microgrid design, operation, and troubleshooting, thereby facilitating community learning about microgrids. If the apprenticeships are developed following certification assessment and training standards such as those set by the Directorate of Industrial Training (DIT) of Uganda, they can enable energy consumers to acquire ISO recognized training (Bogere, 2024, pp. 13, 130).

Community radios, broadcasting in local languages, is another essential medium for spreading awareness and providing educational content on microgrid technologies. Radio remains an important medium for information dissemination and reception in many rural parts of East Africa. In addition to community radios, energy developers and research centres procure talk shows on regular programming of FM radio stations to train and sensitize masses. Talk show programmes tailored to microgrid operations, energy conservation, productive use of energy, and maintenance procedures, among others, are broadcast to educate both skilled technicians and energy consumers in local communities. By using vernacular languages and interactive formats such as call-in sessions, radio programmes promote greater community engagement and knowledge retention.

Online interactive learning platforms and mobile applications have gained importance as alternative skilling methods. With increasing mobile phone penetration in East Africa, training institutions and consultancy firms have leveraged e-learning modules and instructional videos to reach dispersed communities. Platforms, like the VIMLE as described in section 5.3, provide remote training through interactive multimedia resources, quizzes, and virtual mentorship. The VIMLE platform can enable instructors to upload virtual laboratories and other microgrid-related training content (Asiimwe et al., 2024, p. 248; Bogere et al., 2023, p. 673). Such digital approaches complement traditional training by allowing self-paced learning and convenient access to microgrids-related knowledge content. Similarly, didactic materials on leaflets, posters, video shows, and demonstrations in public places such as markets are commonly used forms of knowledge transfer in rural communities of East Africa (Bogere, 2024, p. 88; Raymond et al., 2022, p. 22).

Despite the various skilling methods and knowledge transfer approaches available, challenges remain in reaching disadvantaged populations in remote and off-grid areas. Limited internet access, poor road networks, financial constraints, and language barriers hinder the effectiveness of digital and analogous training programmes. Addressing these issues requires a combination of localized approaches, investment in infrastructure, and policy support to create an inclusive learning ecosystem that benefits disadvantaged communities as well.

In conclusion, skilling community inhabitants in East Africa for microgrid development necessitates a multifaceted approach, incorporating face-to-face training underscoring practical hands-on experiments, demonstrations, and excursions to live microgrids, as well as peer-to-peer learning, on-the-job training, radio broadcasting services, online digital platforms, and other media such as leaflets and posters. By leveraging diverse media and methods, stakeholders can ensure that energy users in local communities are equipped with the necessary skills for microgrid projects' sustainability, ultimately driving energy access and economic empowerment in the East African region.

5.4.3 Communities as a Resource for Learning

The concept of community in the sense of a group sharing common interests, purposes, or values is one that is focused on in community learning. Falk and Harrison (1998) further note that relations and exchange are a focus in the community learning approach. Little progress can be made and then sustained without the oil of the community mechanisms being put in place to aid the smooth running of socio-cultural, economic, and environmental outcomes (Falk & Harrison, 1998, pp. 611, 614). Coudel (2011) reinforces this notion, stating that in order to act, gaining collective awareness of the situation becomes more important than acquiring new knowledge. By focusing on exchange rather than on methods and practices, these approaches often lead to a change in group values through double-loop learning, which enables the group to find new ways of solving problems (Coudel et al., 2011, p. 126). Understanding the community and its ways facilitates sustained and embraced learning. The relevance of community participation in society's transitions is further acknowledged by Wüstenhagen et al. who write that those forms of technological cooperation in which active participants are from various sectors and interact continuously throughout the process are most effective in eliciting social acceptance of renewable energy innovations (2007, p. 2689).

In »Negotiating power acceptance for sustainable minigrid implementation« (Mangeni 2024), the results from the study of East African microgrid communities stress that it is critical that beneficiary communities are custodians of their structures, beliefs, and ways. When microgrid developers and stakeholders bring the idea and practice of the renewable energy projects to the beneficiaries, they interface with the communities. This opens the microgrid developers, operators, regulatory authorities, and all outsiders a field to learn and insight into the community context. The community is key in this learning process. Through participation in baseline assessments, needs assessments, and knowledge exchange processes, the communities hold the outsiders' hands and navigate them through the intricacies of the community at a tempo that draws in the community and enables the harmonious and mutual achievement of the intended energy transformation in a way that the community can and will want to sustain. This brings together contextualization, the project planning of microgrids' developers, and the community involvement. It strengthens the processes of knowledge creation and exchange where all stakeholders are involved. A participatory method, in which all stakeholders are involved, especially the transparently accessed and endorsed community administration structures, contributes to the sustainability of measures and continuity of information (Mangeni, 2024, pp. 176–177).

Conflicts arise when a project approaches a community in a narrow way, without opening up to the diversity of vision, expectations, and possibilities that a new technology brings with it. Central to gaining a deep understanding of a community in order to prevent conflicts is the issue of trust. Trust—among stakeholders, institutions, and society—enables the community to work collectively, consensually, and effectively towards a common goal (Alvial-Palavicino et al., 2011, p. 315). Asiimwe et al. (2024) discuss the variety and discourses of stakeholder conflicts that arise within the learning and implementation processes of the renewable energy projects. The studied microgrid communities revealed that often it is the different stakeholders that direct information according to their roles in the project and thus become media themselves. General solutions may not work and specific situations call for situational solutions in very informal settings. Interlinking the microgrid operator's formal structures with the existing community's formal and informal structures triggers transformational adult learning in situations of energy change. When conflicts occur, they often indicate good reasons for non-acceptance. The level of media and learning acceptance seems to decrease with the level of formality of learning. If community members cannot accept socio-technological change, they risk losing their willingness to being approached by formal learning. This is when artificial media lose their meaning and people must take over (Asiimwe et al., 2024, pp. 249–253).

Further studies of the East African renewable energy terrain, stemming from the A:RT-D Grids interdisciplinary project, show the vital role of the community in community learning in energy transitions. Community guides the diagnosis and informing technical staff of the root causes of technical problems, and adopting technical solutions (Mwammenywa, 2024). It feeds into the contextualizing of learning goals, materials, and modalities (Bogere, 2024). It exposes existing gender discourses (Asiimwe, 2023) and guides conflict-sensitive project implementation (Mangeni, 2024). The roles of the community cannot be overscored enough. They are the cogs and wheels that create a vehicle for community learning and ultimately project sustainability.

5.4.4 Gender Discourses and Community Learning: Challenges and Pathways to Inclusive Participation

Community learning is a cornerstone of SD, particularly in rural regions where formal education systems and resources are often limited. In East Africa, such initiatives are crucial for introducing concepts like renewable energy, which can transform livelihoods and enhance resilience against challenges emanating from energy scarcity. However, gender discourses—deeply rooted societal norms and power dynamics—often create significant barriers to equitable participation in community learning. Addressing these challenges requires a nuanced understanding of the barriers faced by men and women and the development of inclusive strategies that foster the full engagement of all.

In many rural East African communities, traditional gender roles assign men as primary decision-makers and women as caretakers. This division restricts women's access to spaces of public learning and decision-making, often framing their contributions as secondary. As Smith (2020, p. 45) notes, cultural expectations tend to marginalize women's voices in community initiatives, including those focused on SD. For example, women's attendance at community meetings may be perceived as neglecting their domestic duties, leading to societal disapproval. Consequently, their perspectives, which are crucial for addressing community-wide challenges, remain underrepresented.

Low literacy levels further exacerbate women's exclusion from community learning initiatives. According to Nyang'oro (2019, p. 32), gender disparities in education are entrenched in many rural areas due to historical biases that prioritized boys' education over girls. Without access to formal education, many women struggle to grasp technical concepts such as those related to renewable energy technologies. This educational gap not only limits their participation but also perpetuates a cycle of dependency on male counterparts, who are often more literate and better positioned to access resources and information. In addition, economic barriers present another formidable challenge. Women in rural East Africa typically have limited financial resources compared to men due to unequal access to income-generating opportunities. This financial disparity restricts their ability to attend community learning sessions, which may require travel or other expenses. Mutua (2021, p. 94) highlights that women's economic vulnerabilities often force them to prioritize immediate survival needs over long-term learning opportunities. For instance, a woman who spends her day fetching water or farming for subsistence is less likely to allocate time for community learning on renewable energy, even if such knowledge could ultimately benefit her household.

As Kabeer (2020) notes, another critical barrier lies in women's exclusion from decision-making processes. Traditional patriarchal systems often position men as the default decision-makers in families and communities, side-lining women from planning and implementing learning initiatives. This exclusion not only limits the relevance of such sessions to women's lived experiences but also undermines their confidence to participate actively. Kabeer (2020, p. 18) further observes that when women are consistently excluded from decision-making spaces, they internalize these societal constraints, which further perpetuates their marginalization. Yet, logistical and infrastructural challenges also promote a gender wage in community learning. For instance, sessions are frequently scheduled at times or held in locations that are incompatible with women's domestic responsibilities. For instance, a workshop planned in the evening or far from the village centre might exclude women who are burdened with cooking, childcare, or lack of transport. Okello (2017, p. 52) emphasizes the importance of considering logistical barriers when organizing learning sessions, as they disproportionately affect women's ability to attend and engage.

To address these challenges and ensure the inclusivity of renewable energy learning sessions, several strategies can be employed. One foundational step is to conduct gender-sensitive needs assessments before launching any initiative. By engaging both men and women in identifying their unique needs, preferences, and constraints, facilitators can design sessions that resonate with the lived realities of all participants. Smith (2020, p. 48) underscores that inclusive planning not only enhances the relevance of learning initiatives but also fosters a sense of ownership among community members. Flexibility in scheduling is another crucial consideration. According to Mugisha (2018), learning sessions should be timed to accommodate women's domestic responsibilities, such as scheduling meetings during late mornings or early afternoons when household demands are less intense. Mugisha (2018, p. 69) further suggests that adjusting schedules based on community input can significantly increase women's participation. Additionally, providing childcare services during sessions can alleviate one of the primary barriers that prevent women from attending.

Building women's confidence and leadership capacity is essential for fostering active engagement. Facilitators should create a supportive environment that encourages women to voice their opinions and ensures that their contributions are valued. Training women as facilitators can further empower them while inspiring other women to participate. Nyang'oro (2019, p. 34) highlights the importance of representation, noting that women are more likely to engage when they see others like them in leadership roles. Okello (2017, p. 54) points out that simplifying technical concepts not only makes them accessible but also helps demystify topics like renewable energy, which might otherwise seem intimidating especially for women (and men) that are well literate. Engaging community leaders and champions can help shift societal norms that discourage women's participation. Traditional and religious leaders who advocate for gender equality can influence broader acceptance of women's involvement in public learning initiatives. Kabeer (2020, p. 20) notes that leveraging influential voices within the community is a powerful way to challenge entrenched gender biases and encourage behavioural change.

As studies by Mutua (2021) indicate, technology and media offer additional avenues for inclusive learning. Mobile phones and community radio programmes can disseminate renewable energy knowledge to women who are unable to attend in-person sessions. As Mutua (2021, p. 97) observes, such technologies can complement traditional learning formats and ensure continuity in knowledge sharing, particularly in remote areas. Regular assessments of participation rates and feedback from both men and women can help identify barriers and refine approaches. Collecting gender-disaggregated data allows facilitators to track progress and ensure that learning initiatives remain equitable and responsive to community needs (Nyang'oro, 2019, p. 36).

In summary, addressing gender discourses and barriers in community learning is essential for fostering inclusive development in rural East Africa. While challenges such as cultural norms, literacy gaps, economic constraints, and logistical barriers persist, targeted strategies can promote the full participation of both men and women in renewable energy initiatives. By adopting gender-sensitive practices, engaging community stakeholders, and leveraging technology, community learning can become a powerful tool for achieving SD and gender equality in the long run.

5.4.5 References

Alvial-Palavicino, C., Garrido-Echeverría, N., Jiménez-Estévez, G., Reyes, L., & Palma-Behnke, R. (2011). A methodology for community engagement in the introduction of renewable based smart microgrid. *Energy for Sustainable Development, 15*(3), 314–323. https://doi.org/10.1016/j.esd.2011.06.007

Amollo, R. (2021). *Sustainable Solutions for Rural Development in East Africa.* Progress Press.

Asiimwe, H., Bode, H., Bogere, P., Freitag, C., & Mangeni, T. (2024). Which media for whom? the implementation of microgrids as a trigger of transformational adult learning opportunities in formal, informal and situational settings in times of change. In E. Andrzejewska, E. Matthes, S. Schütze, & J. Van Wiele (Eds), *Educational Media for Adults* (pp. 245–255). Verlag Julius Klinkhardt. https://doi.org/doi.org/10.35468/6126-18

Bogere, P. (2024). *Knowledge Transfer for Microgrids Sustainability: The Case of East Africa.* Doctoral Thesis, Paderborn University. https://doi.org/10.17619/UNIPB/1-2167

Bogere, P., Bode, H., & Temmen, K. (2023). Work-in-progress: development of a virtual and interactive microgrids learning environment for microgrids sustainability—the case of East Africa. In M. E. Auer, W. Pachatz, & T. Rüütmann (Eds), *Learning in the Age of Digital and Green Transition. ICL 2022. Lecture Notes in Networks and Systems* (Vol. 1, pp. 671–679). Springer Nature.

Bogere, P., Temmen, K., & Bode, H. (2023). Knowledge transfer concepts for microgrids sustainability—the case of East Africa. *IEEE Global Engineering Education Conference (EDUCON)*, May, 1–6. https://doi.org/10.1109/EDUCON54358.2023.10125208

Coudel, E., Tonneau, J. P., & Rey-Valette, H. (2011). Diverse approaches to learning in rural and development studies: review of the literature from the perspective of action learning. *Knowledge Management Research and Practice, 9*(2), 120–135. https://doi.org/10.1057/kmrp.2011.12

Falk, I., & Harrison, L. (1998). Community learning and social capital: »just having a little chat«. *Journal of Vocational Education and Training, 50*(4), 609–627. https://doi.org/10.1080/13636829800200063

Heiskanen, E., Nissilä, H., & Tainio, P. (2017). Promoting residential renewable energy via peer-to-peer learning. *Applied Environmental Education and Communication, 16*(2), 105–116. https://doi.org/10.1080/1533015X.2017.1304838

Johnson, T. (2019). Women and renewable energy: overcoming challenges in Africa. *Energy Policy Journal, 11*(4), 121–128.

Kabeer, N. (2020). *Gender, Development, and Learning: A Social Inclusion Perspective.* Oxford University Press.

Kaneza, P., Nkurunziza, J. B., & Twagilimana, I. (2024). Effect of hands-on laboratory experiment and demonstration teaching techniques on Rwandan secondary school students' academic performance in solutions and titration. *African Journal of Research in Mathematics, Science and Technology Education, 28*(1), 44–56. https://doi.org/10.1080/18117295.2023.2275858

KFW Development Bank, GIZ, & IRENA. (2020). *The Renewable Energy Transition in Africa: Powering Access, Resilience and Prosperity.*

Krüger, J. T., Höffler, T. N., Wahl, M., Knickmeier, K., & Parchmann, I. (2022). Two comparative studies of computer simulations and experiments as learning tools in school and out-of-school education. *Instructional Science, 50*(2), 169–197. https://doi.org/10.1007/s11251-021-09566-1

Mangeni, T. N. (2024). *Negotiating Power Acceptance for Sustainable Minigrid Project Implementation.* Universität Bibliothek Paderborn. https://doi.org/10.17619/UNIPB/1-2144

Mugisha, J. (2018). Barriers to women's participation in community learning initiatives in East Africa. *African Journal of Development Studies, 12*(2), 65–70.

Mutua, C. (2021). *Economic Empowerment and Gender Equality in Rural Africa.* Green Leaf Press.

Mwakisisya, H. J. (2020). Meeting on-job adult learners' special needs: insights from improved opportunities and obstacles to development (o and od) system's trainings in Tanzania. *Journal of Education & Social Policy, 7*(1). https://doi.org/10.30845/jesp.v7n1p22

Mwammenywa, I. (2024). *A Novel Autonomous and Real-time Load Monitoring and Control System in Microgrids based on Fuzzy Logic Control and LoRa Wireless Communication.* Doctoral Thesis, University of Paderborn. https://doi.org/10.17619/UNIPB/1-2019

Mwangi, K. (2018). The role of gender in community-based energy projects. *Journal of Energy Research in Africa, 6*(1), 15–22.

Nyang'oro, P. (2019). Educational challenges for women in rural Africa. *Journal of Education and Development, 25*(3), 30–37.

Okello, A. (2017). *Bridging Gender Gaps in Community Development Projects.* East African Development Press.

Otieno, L. (2020). *Inclusive Learning Models for Developing Communities.* Horizon Books.

Raymond, J., Kihedu, J. H., & Kimambo, C. Z. M. (2022). A model for sustainable adoption of solar photovoltaic technology in Tanzania. *Tanzania Journal of Engineering and Technology, 41*(2), 16–34. https://doi.org/10.52339/tjet.v41i2.776

Smith, R. (2020). *Community Learning in Developing Countries: Lessons from the Field.* Routledge.

Wüstenhagen, R., Wolsink, M., & Bürer, M. J. (2007). Social acceptance of renewable energy innovation: an introduction to the concept. *Energy Policy, 35*(5), 2683–2691. https://doi.org/10.1016/j.enpol.2006.12.001

Christine Freitag, Katrin Temmen

6. Prospects: Knowledge Distribution and Educational Formats in Years to Come

This book has been a joint (ad)venture for those involved in its planning, writing, editing, and coordinating. It presents the product of our efforts, as described in our introduction. Reading all parts of the book now assures us that it is what we wanted it to be: the second pillar of our A:RT-D Grids project, that is, to transform years of research into information and teaching. This is why we are not just talking about research carried out and analysed, but also its applicability to East African locations, situations, and communities.

In our empirical data, we found many hints that the introduction of sustainable energy into local communities is often accompanied by so-called educational measures (see chapter 5), which often turn out to be leaflets or pieces of information given by individuals. We believe to have shown in this book that sustainable energy implementation needs more, and that different target groups have to be addressed differently to make change possible. Not every person consuming electricity from a community microgrid will need to know details of microgrid design and possibilities for improvement. Equally, not every technician will want to know all the details of community learning or gender discourses. None of us would expect every reader to have read the whole book, but we would be delighted if it had opened up new perspectives, uncovered previously unknown aspects of well-known practices, and especially made discoveries that will help readers to understand why things have (not) worked in the past. One should allow for encounters with other perspectives, as we intended to do ourselves, between the technical and social sides of sustainable energy transformation.

We, as authors and editors, want to utilize the book for more discussion and all sorts of workshops, especially in collaboration with enterprises, communities, schools, adult education centres, and universities. Some such events have already taken shape; some are in the planning stages; others need to be designed and facilitated. On a lower and yet broader level, we would have liked to offer a new format of virtual exchange. For now, we would be happy if our initiative could foster exchange in existing social media networks and groups, such as those on *LinkedIn* or *facebook*, among others, which bring together people on different aspects of our topic, like the *Solar Energy Network*, the *Africa Minigrid Developers Association*, *The Global Energy Alliance for People and Planet*, or *UNESCO's ESD-net* for those interested in Education for Sustainable Development around the world (all of these meet on *LinkedIn*), or groups like that of *environment and sustainability professionals* or on the *UN SDGs* (these meet on *facebook*). Of course, there are many more (virtual) places where we can meet and discuss what we share. Let us stay on track and keep up the momentum.

List of Figures

Fig. 1 Lights kept on during daytime in the Tanzanian village—many were observed during data collection (reproduced from Bogere, 2024, p. 67). 34

Fig. 2 Light and socket circuits terminated in the same JB and using the same cable size—1.5 mm^2 (Bogere, 2024, p. 69). 35

Fig. 3 The African map showing the three national states of Tanzania, Uganda, and Kenya, which were the focus of this study. Source: Danny (2004). 42

Fig. 4 Solar irradiance values (global irradiance, horizontal) of Africa (World Bank Group / SolarGIS 2019). 47

Fig. 5 Wind map of Sub-Saharan Africa: Average wind speeds. (World Bank Group, Global Wind Atlas, DTU 2019). 49

Fig. 6 Geothermal heat flow intensity in Africa. Source: Al-Aghbary et al. (2022). 50

Fig. 7 Percentage composition of the total electricity mix of the grid installed capacity in Tanzania in 2023 (Tanzania Power Production and Demand—2024 Update). 52

Fig. 8 Uganda's electricity generation capacity (in MW) together with relative share of overall capacity (in %) as of 2022. (ERA, 2023). 52

Fig. 9 Uganda's electricity generation from grid-connected power plants in 2023 (ERA, 2024). 53

Fig. 10 Kenya's installed electricity capacity (in MW) and relative share of total capacity (in %) as at 2023 (Kenya National Bureau of Statistics, 2024). 54

Fig. 11 Relative shares in Kenya electricity generation of total 13.3 TWh over the year ended June 2023 (EPRA, 2023). 54

Fig. 12 Load profile of the three phases in Mpale microgrid in Tanzania (Philipo et al., 2023). 56

Fig. 13 Average load profile for community microgrid for 15–29 December 2022 (Kakande et al., 2024). 57

Fig. 14 Load profile for Senyondo microgrid from March 2022 to September 2023 (the dashed line represents the average values) (Kakande et al., 2024). 57

Fig. 15 General circuit symbol of a converter. 71

Fig. 16 Wording for power electronic converters. 72

Fig. 17 Basic switch element of a DC-DC converter consisting of a transistor and a diode. 74

Fig. 18 Circuit diagram of a single-phase PV inverter (shown with a grid connection). 74

Fig. 19 Principle of Pulse Width Modulation (PWM) for a PV inverter. Plot 1 (upper): Desired sinusoidal grid voltage and the so-called triangular-shaped modulation carriers. Plots 2 & 3: The switching commands for the two inverter legs, resulting from comparison of grid voltage and the respective modulation carrier. Plot 4 (lowest): The »chopped« inverter output voltage in comparison with ideal sinusoidal curve. 75

Fig. 20 Circuit diagram of the LWAK microgrid system. 77

Fig. 21 Interconnection plan between the three different subsystems (A, B, and C) of the convent. 84

Fig. 22 Preparation of the roof for the PV generator in Lwak after delivery of the PV modules. 89

Fig. 23 Installation of the rooftop PV generator in Lwak. 90

Fig. 24 Demand-side management techniques (adapted from Lampropoulos, 2014). 96

Fig. 25 Site layout in Lwak, Kenya (Kakande et al., 2025). 97

Fig. 26 Communication tools to carry out DSM in Lwak (Kakande et al., 2025). 98

Fig. 27 Energy drawn by appliances with and without DSM (Kakande et al., 2025). 99

Fig. 28 Flow chart of the cooling appliances' control mechanism of switching ON and OFF appliances based on temperature. 100

Fig. 29 Daily profile before and after temperature-based control for Freezer 8. 101

Fig. 30 Daily profile before and after temperature-based control for Fridge 4. 102

Fig. 31 Role of communication in attaining a sustainable microgrid (Tightiz & Yang, 2020). 106

Fig. 32 Community microgrid schematic diagram of case study (Mwammenywa, 2024). 109

Fig. 33 Decentralized load monitoring and control architecture in the microgrid (Mwammenywa, 2024). 110

Fig. 34 Scheme of power line monitoring and supply control for customers (Mwammenywa, 2024). 111

Fig. 35 Flow chart for demand response and load management (Mwammenywa, 2024). 112

Fig. 36 Energy harvesting from AC power line (Mwammenywa, 2024). 112

Fig. 37 Bidirectional data communication via LoRa transceivers (Mwammenywa et al., 2022). 113

Fig. 38 FIS-Based Model for Load Control Signal (Mwammenywa, 2024). 114

Fig. 39 Correlation Between Inputs (PV Generation and Battery SoC) and Output (Total Load Limit) (Mwammenywa, 2024). 114

Fig. 40 Flow of Information from Installed Module Through LoRa Transceiver to Host Computer (Mwammenywa, 2024). 115

Fig. 41 Updated Schematic Diagram of the Community Microgrid (Mwammenywa, 2024). 116

Fig. 42 Example for indication by »traffic light«: Yellow light signalling to alert a consumer of usage being close to a load limiting threshold (Mwammenywa, 2024). 117

Fig. 43 Factors influencing willingness to pay showing the mediating perceptions. 130

Fig. 44 Reinforcing cycle of hierarchical governance and technical dependence in microgrid management. 148

Fig. 45 The Democratized Technical Training Framework (DTTF). 150

Fig. 46 Features of the Tanzanian microgrid (Ngoti, 2024a; 2024b). 155

Fig. 47 Governance influence distribution in the community microgrid. 165

Fig. 48 Hierarchical clustering of governance influence. 166

Fig. 49 Governance influence heatmap. 167

Fig. 50 TLS smart metering architecture (Namujju et al., 2024, p. 2). 182

Fig. 51 The conflict triangle (Galtung, 1996, p. 72). 191

Fig. 52 Adaptation by Mangeni 2024 using field data. Diana Francis, Guus Meyer (Fisher et al., 2000, p. 121). 205

Fig. 53 Appliances and components for microgrid-related training at Don Bosco Oysterbay VTC in Dar es Salaam, Tanzania (reproduced from Bogere, 2024, p. 89). Left: GT participants tasked with modelling a microgrid demonstrator. Right: Assorted appliances for microgrids training. 239

Fig. 54 A pictorial summary of the microgrids course lesson during which learners were tasked with wiring a microgrid demonstrator (photos reproduced from Belov, 2023, pp. 47–64). 240

Fig. 55 A CA frame for selected ILOs (adapted from Bogere et al., 2023, p. 676). 249

Fig. 56 VIMLE core system design based on the MVC architecture (reproduced from Bogere, 2024, p. 118; originally adapted from Pandit et al., 2010, p. 57). 251

Fig. 57 Illustration of the VIMLE architectural design based on WebRTC technology (reproduced from Bogere, 2024, p. 116; Bogere et al., 2023, p. 674). 253

Fig. 58 VIMLE architectural design illustration indicating selected VIMLE service MSs, local and remote clients (reproduced from Bogere, 2024, p. 120). 254

Fig. 59 VIMLE system layout (reproduced from Bogere, 2024, p. 122; originally adapted from Bogere et al., 2023, p. 677). 255

List of Tables

Tab. 1 Hydropower plants in Tanzania (operational, planned, and ongoing) 45

Tab. 2 Average daily refrigeration power draw with and without DSM (Kakande et al., 2025) 101

Tab. 3 Comparison of communication technologies in the context of micro-grids (Serban et al., 2020) 111

Tab. 4 Access and generation of electricity in East Africa (author's own compilation) 124

Tab. 5 The IAD framework of microgrids 127

Tab. 6 Descriptive statistics 139

Tab. 7 Binary Logistic Average Marginal Effects (AME) for WTP > 10,000 141

Tab. 8 Case Study Microgrid—Survey themes and prevailing stakeholder sentiment 147

Tab. 9 Actor-level Influence 173

Editors and Authors

Editors

Prof. Dr. Christine Freitag (*1964) is professor of education science at Paderborn University. Her main focus is on comparative education, linked to different perspectives on Education for Sustainable Development.

Prof. Dr. Katrin Temmen (*1969) is professor of didactics of technology at Paderborn University. Her main focus is on innovative didactic concepts for engineering courses and technical teacher training.

Prof. Dr. Burkhard Hehenkamp (*1969) is professor of institutional economics and economic policy at Paderborn University. His research centers on institutional design in various areas of application such as health care markets, sustainability, and industrial organization.

Prof. Dr. Stefan Krauter (*1963) works on sustainable energy supply. His Ph.D., finalized in 1993, dealt with accurate yield prediction of solar PV. From 1996 until 2006 he lectured as a professor at the Federal University of Rio de Janeiro and the State University of Ceara, both in Brazil. In 2006, he co-founded Photovoltaic Institute Berlin, during 2008–2010 he served as Professor at the University of Applied Sciences in Biberach, since 2010 he is a Professor at Paderborn University, filling the chair of Electrical Energy Technologies.

Tobias Klaus (*1967) is a political scientist by training and works as Director of International Projects and Research at the ECOLOG Institute for Social-Ecological Research and Education in Lüneburg. He has also implemented a number of rural electrification and infrastructure projects in several African and Asian countries.

Other Authors

Henrietta Acquah-Swanzy
Dr. Henry Asiimwe
Prof. Dr. Joachim Böcker
Henrik Bode
Dr. Paul Bogere
Prof. Dr. Ulrich Hilleringmann
Josephine Nakato Kakande
Dr. Teddy Mangeni
Dr. James Musana
Dr. Ibrahim Mwammenywa
Irene F. Ngoti
Dr. Lillian Donna Namujju
Dr. Godiana Philipo
Stefan Trittler